Phase Transformations and Heat Treatments of Steels

Phase Transformations and Heat Treatments of Steels

Bankim Chandra Ray, Rajesh Kumar Prusty, and Deepak Nayak

CRC Press is an imprint of the
Taylor & Francis Group, an **informa** business

First edition published 2020
by CRC Press
6000 Broken Sound Parkway NW, Suite 300, Boca Raton, FL 33487-2742

and by CRC Press
2 Park Square, Milton Park, Abingdon, Oxon, OX14 4RN

© 2020 Taylor & Francis Group, LLC

CRC Press is an imprint of Taylor & Francis Group, LLC

Reasonable efforts have been made to publish reliable data and information, but the author and publisher cannot assume responsibility for the validity of all materials or the consequences of their use. The authors and publishers have attempted to trace the copyright holders of all material reproduced in this publication and apologize to copyright holders if permission to publish in this form has not been obtained. If any copyright material has not been acknowledged, please write and let us know so we may rectify in any future reprint.

Except as permitted under U.S. Copyright Law, no part of this book may be reprinted, reproduced, transmitted, or utilized in any form by any electronic, mechanical, or other means, now known or hereafter invented, including photocopying, microfilming, and recording, or in any information storage or retrieval system, without written permission from the publishers.

For permission to photocopy or use material electronically from this work, access www.copyright. com or contact the Copyright Clearance Center, Inc. (CCC), 222 Rosewood Drive, Danvers, MA 01923, 978-750-8400. For works that are not available on CCC, please contact mpkbookspermissions@ tandf.co.uk

Trademark notice: Product or corporate names may be trademarks or registered trademarks and are used only for identification and explanation without intent to infringe.

ISBN: 978-0-367-02868-8 (hbk)
ISBN: 978-0-429-01921-0 (ebk)

Typeset in Times
by codeMantra

Contents

Preface...ix
Authors...xi

Chapter 1   An Introduction to Metals 1

      1.1   Elements, Atoms, and Isotopes 1
      1.2   Types of Bonding between Atoms..........................2
      1.3   Crystal Structures 5
      Further Reading..19

Chapter 2   Diffusion..21

      2.1   Atomic Diffusion Mechanisms 21
      2.2   Types of Diffusion 22
      Further Reading..36

Chapter 3   Defects in Crystalline Solids...37

      3.1   Introduction 37
      3.2   Classification 37
      Reference...50
      Further Reading..50

Chapter 4   Solid Solutions...51

      4.1   Introduction 51
      4.2   Types of Solid Solutions 51
      4.3   Electron-to-Atom Ratio 53
      4.4   Enthalpy of Formation of a Solid Solution 54
      4.5   Entropy of Formation of a Solid Solution..............56
      4.6   Free Energy Change upon Formation of a Solid Solution.......58
      4.7   Ordered and Random Solid Solutions 60
      4.8   Intermediate Phases...61
      Further Reading..62

Chapter 5   Phase Diagrams and Phase Transformations 63

      5.1   Thermodynamic Considerations of Phase Diagrams.............63
      5.2   Gibb's Phase Rule 64
      5.3   Lever Rule 65
      5.4   Types of Phase Diagrams and Phase Transformations...........66
      5.5   Some Other Solid-Phase Transformations in
               Metals and Alloys..76

vi   Contents

| | 5.6 | Roles of Defects and Diffusion | 77 |
| | | Further Reading | 78 |

Chapter 6 Iron–Carbon Phase Diagram 79

	6.1	Introduction	79
	6.2	Allotropic Transformations in Iron	80
	6.3	Solubility of Carbon in Iron	80
	6.4	Iron–Iron Carbide Phase Diagram	82
	6.5	Effect of Alloying Elements on the Iron–Carbon Equilibrium Diagram	86
		Further Reading	90

Chapter 7 Thermodynamics and Kinetics of Solid-State Phase Transformation 91

	7.1	Introduction	91
	7.2	Nucleation	91
	7.3	Growth Kinetics	98
	7.4	Time–Temperature Transformation and Continuous Cooling Transformation Diagrams	102
		References	106
		Further Reading	106

Chapter 8 Phase Transformation in Steels 107

	8.1	Introduction	107
	8.2	Formation of Austenite	107
	8.3	Pearlitic Transformation	108
	8.4	Bainitic Transformation	111
	8.5	Martensitic Transformation	113
		Further Reading	116

Chapter 9 Heat Treatment Furnaces 117

	9.1	Introduction	117
	9.2	Classification of Furnaces	117
	9.3	Batch Furnace	118
	9.4	Continuous Furnace	123
	9.5	Salt Bath Furnace	125
		Further Readings	125

Chapter 10 Heat Treatment Atmosphere 127

	10.1	Introduction	127
	10.2	Reactions between Atmosphere and Material	127
	10.3	Types of Furnace Atmospheres	129
		Further Reading	131

Contents vii

Chapter 11 Common Heat Treatment Practices.. 133

    11.1   Introduction 133
    11.2   Typical Heat Treatment Processes....................................... 133
    11.3   Hardenability 153
    11.4   Case Hardening and Surface Hardening 158
    11.5   Thermomechanical Treatment.. 164
    11.6   Heat Treatment of Carbon and Alloy Steels......................... 166
    Further Reading 168

Chapter 12 Special Steels.. 169

    12.1   Stainless Steels 169
    12.2   Hadfield Manganese Steels.. 172
    12.3   High-Strength Low-Alloy or Microalloyed Steels 172
    12.4   Transformation-Induced Plasticity Steels............................ 172
    12.5   Maraging Steels 173
    12.6   Dual-Phase Steels 173
    12.7   Tool Steels 173
    12.8   Electric Grade Steels 177
    Further Reading 177

Chapter 13 Some *In Situ* Postweld Heat Treatment Practices 179

    13.1   Necessity.. 179
    13.2   Conventional Postweld Heat Treatment Process 179
    13.3   *In Situ* Postweld Heat Treatment of Transformation-
           Induced Plasticity Steel 180
    13.4   Postweld Heat Treatment of Duplex Stainless Steel............. 182
    References 184

Chapter 14 Heat Treatment of Cast Iron 185

    14.1   Introduction 185
    14.2   Types of Cast Iron 185
    14.3   Heat Treatment of Gray Cast Iron 192
    14.4   Heat Treatment of Malleable Cast Iron 193
    14.5   Heat Treatment of Spheroidal Graphite Irons 194
    Reference.. 196
    Further Reading 196

Chapter 15 Heat Treatment Defects and Their Determination............................ 197

    15.1   Distortion... 198
    15.2   Warping 198
    15.3   Residual Stresses 199
    15.4   Quench Cracking.. 199

viii   Contents

15.5	Soft Spots	200
15.6	Oxidation and Decarburization	201
15.7	Low Hardness and Strength after Hardening	202
15.8	Overheating of Steel	203
15.9	Burning of Steel	204
15.10	Black Fracture	205
15.11	Deformation and Volume Changes after Hardening	205
15.12	Excessive Hardness after Tempering	205
15.13	Corrosion and Erosion	206
	References	206
	Further Reading	206

Chapter 16 Some Special Heat Treatment Practices.................................207

16.1	Automobile Industries	207
16.2	Aerospace Industries	210
16.3	Medical Equipment	214
16.4	Defense Industries	216
	References	217

Index...219

Preface

The study of phase transformations, in terms of subtly underlying thermodynamics and kinetics with the probability of attachment of atoms to critical nuclei, forms the scientific aspect of formation of microstructures, and subsequently, the practical practices of phase transformations are commercially achieved through heat treatments. As a matter of fact, heat treatment solely involves managing the kinetics of phase transformation to achieve desirable properties required in technological applications. This book explains mechanisms, thermodynamics, and kinetics of phase transitions in materials engineering field of steels and provides concepts to commercialization-oriented understanding of heat treatment practices and principles on the knowledge of phase transformations. Phase transformation and heat treatment are indeed connected concurrently and, in the authors' opinion, should be treated as complementary subjects as opposed to distinct subjects. *Phase Transformations and Heat Treatments of Steels* attempts to fill this gap, prevalent in the available literature. The book's unique presentation links basic understanding of theory with application in a perpetually progressive yet exciting manner.

The book focuses on the processing–structure–properties triangle/structure–property correlation, as it introduces the fundamental principles of physical metallurgy, phase transformations, and heat treatments of steel legacy as a structural material. It supplies a broad overview of specific types of phase transformations, supplemented by practical case studies of engineering alloys. The lucid, well-organized, and eminently readable text not only provides a proficient analysis of all the relevant topics but also makes them comprehensible to the readers through the adept use of numerous diagrams and illustrations. The book takes a pedagogical approach and analyzes all concepts systematically and logically, making it ideal for those new to the field.

While attempting to provide a comprehensive understanding of phase transformations with the practice of heat treatment and their interrelationships, the authors have tried to provide insights into specialized processes and practices, and to convey the excitement of the atmosphere in which new and different properties are introduced and tailored.

Divided into 16 chapters, the material is organized in a logical progression, beginning with fundamental principles and then building to more complex concepts involved in phase transformation and heat treatment of almost all types of steels as well as cast irons.

The book begins with a clear exposition of the basic concepts and definitions related to metals, chemical bonding, and the structure of solids in Chapter 1. Chapters 2 and 3 provide a theoretical background to solid-state diffusion and crystal imperfections. These chapters also provide a deep insight into the structural control necessary for optimizing the various properties of materials. Then, a detailed discussion on solid solutions with their types and the associated thermodynamics follows in Chapter 4. The first part of Chapter 5 offers an in-depth analysis of phase diagrams, while the latter part covers the structure and change of structure through phase transformations. Chapter 6 discusses, in rich detail, the iron–carbon phase diagram, which is indispensable in the field of metallurgy and materials engineering. Chapter 7 emphasizes the thermodynamic and kinetic aspects of solid-state phase transformation. This includes a comprehensive discussion on nucleation and growth in the context of phase transformations in general. Chapter 8 extends the theory of phase transformations to steels. Chapter 9 provides succinct summaries of common furnaces used for heat treatment. This is followed by a thorough discussion on heat treatment atmosphere in Chapter 10. Chapter 11 is devoted to a detailed description of the prevalent heat treatment practices and their purposes. It also highlights the topics of hardenability, hardening treatment, and thermomechanical treatment. Chapter 12 introduces certain special steels in order to pique readers' interest for understanding the current and futuristic aspects of steels application. Chapters 13 and 14 describe *in situ* postweld heat treatment practices and heat treatments of cast irons, respectively. Chapter 15 is dedicated to discussing defects produced due to heat treatment and their characterization. Chapter 16 contributes toward rounding out readers' knowledge regarding certain practical and industrial heat treatment techniques. The 16 chapters of the book are organized for almost linear purview in a graduate-level course, in order to provide concrete understanding through steps. Accordingly, each chapter serves a different purpose, but all chapters are connected to provide an assimilated knowledge of phase transformation through proper correlation between structural evolution and corresponding thermodynamic kinetics.

Designed primarily as an introductory text for undergraduate and postgraduate students of metallurgy, the book also serves the needs of allied scientific disciplines at the undergraduate and graduate levels. With its excellent balance between the fundamentals and advanced information, it can also serve as an invaluable guide for practicing professional engineers and scientists.

The authors are indeed privileged to extend their sincere appreciation and acknowledgment to students, staff, scientists, and faculties of National Institute of Technology, Rourkela, and CSIR-Institute of Minerals and Materials Technology, Bhubaneswar, for their insightful suggestions incorporated into this scholarly manuscript. We shall remain grateful to all around us here for their contributions to meet the need of present students and scholars.

All family members of all authors are truly indebted in our hearts for their unconditional soulful support.

Authors

Bankim Chandra Ray, Phd, has worked at the National Institute of Technology (NIT), Rourkela, India, since 1989. A dedicated academician with more than three decades of experience, Bankim Chandra Ray is a full professor since 2006 at the NIT, Rourkela. He earned his PhD from the Indian Institute of Technology, Kharagpur, India, in 1993. Apart from instructing students in the field of Phase Transformation and Heat Treatment, he has also guided many master's degree and PhD scholars. He has made seminal contributions in the field of Phase Transformation and Heat Treatment and Composite Materials.

An adept administrator, he has also served as the Dean of Faculty, Head of the Department of the Metallurgical and Materials Engineering, and also an incumbent Coordinator of Steel Research Center at the NIT, Rourkela.

His research interests are mainly focused on the mechanical behavior of FRP composites. He is leading the Composite Materials Group at the NIT, Rourkela, a group dedicated to realizing the technical tangibility of FRP composites (https://www.frpclabnitrkl.com). With numerous highly cited publications in prominent international journals, he has contributed extensively to the world literature in the field of material science and engineering. He also holds a patent deriving from his research. With nearly 155 publications in international journals, Prof. Ray also authored many books/book chapters from leading publishers. He has been associated with several prestigious societies such as Indian Institute of Metals, Indian National Academy of Engineering, and several government and private institutes and organizations. As an advisor to New Materials Business, Tata Steel Ltd., he has been instrumental in facilitating the steel magnate's foray into the FRP composites business. His constant endeavors toward academics and his field of specialization have been unparalleled, and yet thoroughly inspiring for many of the young engineering minds.

Rajesh Kumar Prusty, PhD, is an enthusiastic and focused teacher who is committed to safeguarding and promoting the education and well-being of students and young people at all times in a friendly and conducive atmosphere. Presently he is an assistant professor at the Department of Metallurgical and Materials Engineering, NIT, Rourkela, India, from 2014. He has graduated with a master's degree (ME) in Materials Engineering from Indian Institute of Science, Bangalore, and earned a PhD from the NIT, Rourkela. He is fascinated in opening up new dimensions of research particularly in the domain of structural materials with the ambition of making these materials very reliable in the Indian market. He has authored more than 25 SCI/Scopus journal articles and one book related to FRP composite materials, and at present, he is a principal investigator of three ongoing projects sponsored by the DRDO, CSIR, and SERB. He actively contributes to the FRP composite lab at the NIT, Rourkela (https://www.frpclabnitrkl.com/) in terms of visualizing and identifying real-time problems associated with FRP composites followed by conceptualizing and designing experiments and work plan accordingly to address those issues through constructive and logical techniques. He has supervised 16 master's and 16 bachelor's theses up to the present. Currently, he has five doctoral and five master's scholars working on FRP composites.

Deepak Nayak is a scientist at the CSIR-Institute of Minerals and Materials Technology, Bhubaneswar, since 2015. Born in 1989 in Bhubaneswar, he graduated from the NIT, Rourkela, with a BTech in Metallurgical and Materials Engineering before working as a graduate engineer trainee at JSL Stainless Ltd., New Delhi, in 2010. Afterward, he returned to Bhubaneswar, where he earned his MTech from Academy of Scientific & Innovative Research (CSIR-IMMT) in 2013. He manages projects from concept to commercialization in the areas of ferrous metallurgy, including mineral beneficiation of low-grade iron ores, microwave processing, and extraction of valuable metals.

1 An Introduction to Metals

1.1 ELEMENTS, ATOMS, AND ISOTOPES

Chemical elements are the fundamental matters of all the materials. These elements are chemically distinct and exhibit unique physical and mechanical properties. The basic representative block of any element is again the atoms, which is comprised of electrons (negatively charged), protons (positively charged), and neutrons (neutral). Both proton and neutron are of almost similar weight, which is around 1.67×10^{-27} kg, whereas electron is much lighter in weight around 9.11×10^{-31} kg. The weight of an atom is almost the same as that of the nucleus, which contains the neutrons and protons. However, the diameter of an atom ($\sim 10^{-10}$ m or 1 Å) is quite larger than that of the nucleus ($\sim 10^{-14}$ m). The magnitude of charge of an electron and a proton being equal, the atom contains exactly the same number of both the entities in order to maintain electrical neutrality. The atomic number of an element indicates the number of protons it possesses in a single atom which is also the same as the number of electrons. However, the difference between atomic weight and atomic number usually indicates the average number of neutrons in the atom. The periodic table designed by the Russian scientist Mendeleev is the ideal tool to find out these numbers for any specific element. Almost all the empty cells in the original Mendeleev's periodic table have been filled due to discovery of new elements in subsequent time. Some elements may have a higher or lower number of neutrons than that of the electrons/protons. As these elements have the same number of electrons/protons, the atomic number is not changed. However, due to different neutrons, the atomic weight becomes different. These are called isotopes. The most common example of isotopes is hydrogen. A high fraction of hydrogen atoms comprises only one proton without any neutron, thus having an atomic weight of 1. A small fraction of hydrogen atoms contains one proton and one neutron giving rise to an atomic weight of 2, which are most commonly known as deuterium. Another small fraction of hydrogen atoms known as tritium (atomic weight 3) possess one proton and two neutrons. In all these cases, the number of protons is 1, and all these elements have the same atomic number, but with different atomic weight. Most of the elements in nature are mixture of such multiple isotopes, and thus, the atomic mass is not always a whole number and is the weighted average of the atomic weights of these isotopes. Continuing with the example of hydrogen, in commonly available hydrogen, the isotopes are mixed in such a proportion that the average atomic weight is 1.008. Taking the examples of iron (Fe), it is naturally available in the form of four stable isotopes. The most abundantly available form of iron is ^{56}Fe (~91.754%) followed by ^{54}Fe (~5.845%), ^{57}Fe (2.119%), and ^{58}Fe (0.282%), giving rise to an average atomic mass of 55.85.

1.1.1 Types of Elements

Based on the broad physical and mechanical properties of elements, they are categorized into three groups, i.e., (i) metals, (ii) metalloids, and (iii) nonmetals. Typically, metals are of shiny lustrous appearance, when prepared fresh or fractured. They are good conductors of heat and electricity. Metals can be plastically deformed and are normally malleable (can be made to thin sheets) and ductile (can be drawn into wires). Except mercury, all other metals remain in their solid state at normal room temperature and exhibit crystalline arrangement of atoms. Around 91 out of 118 elements in the periodic table are metal. However, the exact number is not available, as the boundaries between metals, nonmetals, and metalloids fluctuate due to lack of globally accepted basis of categorization. Metals constitute around 25% of the earth's crust and are inseparable from the present era of civilization. To a large extent, the development of civilization is driven by development in the field of metals and associated products. In the same line, a nonmetal is defined as an element that lacks in the metallic properties. Low density, boiling temperature, and melting temperature are some key physical properties of nonmetals. Most of the nonmetals are gases at room temperature and usually poor conductors of heat and electricity. Some nonmetals are brittle solids at room temperature but good conductors of electricity and heat, e.g., carbon. Metalloids exhibit properties that are in between metals and nonmetals or a mixture of metals and nonmetals. Boron, silicon, germanium, arsenic, antimony, and tellurium are the well-accepted metalloids. They usually have metal-like lustrous appearance but are only moderate conductors of heat and electricity.

1.2 TYPES OF BONDING BETWEEN ATOMS

In general, materials in their solid state exhibit a well-arranged array of atoms forming a regular geometric pattern to reduce their free energy. However, solids such as glass, wax, and paraffin do not follow this trend, and the arrangement of atoms is similar to that of the liquid state and thus termed as amorphous solids. The next important point now is to consider the types of forces responsible to have bonds between adjacent atoms in order to complete the structure of the solid. Generally, there exist four types of interatomic bonding.

1.2.1 Ionic Bond

Ionic bonding is preferable for elements having large difference in electron negativity. All the elements tend to become stable upon achieving octet configuration, i.e., exactly eight electrons on the outermost shell. However, many of the pure elements do not satisfy this criterion. For example, sodium (atomic number 11) is having only one electron in its outermost shell. Hence, the easiest way for sodium to achieve octet configuration is to donate this single electron. However, as the process does not involve transfer of protons, upon donating the single electron, it becomes positively charged and is termed as a positive ion or cation. For sodium, it is denoted as Na^+. All the elements in group IA of the periodic table in this way tend to be single positive cations due to their single valency (e.g., Na^+, K^+, Cs^+).

An Introduction to Metals

In a similar way, group IIA elements become double positively charged upon donating two electrons to acquire their respective octet configuration (e.g., Mg^{2+}, Ca^{2+}, Ba^{2+}). On the contrary, the number of electrons in nonmetals generally is close to eight. Hence, the easiest way for them to achieve the stable octet configuration is to accept the required number of electrons (eight – number of electrons in outermost shell) from somewhere else to have exactly eight electrons at the outermost shell. Considering the case of chlorine, which is having seven electrons in its outermost shell, a single electron is required to be stable. If it gets that electron from somewhere, it becomes single negatively charged (as there is one more electron than the number of protons). This negatively charged ion is termed as anion and denoted as Cl^-. Elements corresponding to group VIIA thus tend to be single negatively charged (e.g., F^-, Cl^-, Br^-). Similarly, elements in group VIA require two electrons to achieve octet configuration and thus become doubly charged anions (e.g., O^{2-}, S^{2-}). Now, this process of obtaining octet configuration is possible only when there is mutual exchange of electrons, such that electrical neutrality is maintained. For example, the electron donated by Na is accepted by the nearby Cl atom in order to neutralize the charge and form a compound named sodium chloride (NaCl) or normal table salt (Figure 1.1). One point here is to note that the properties of the ionic compound formed have nothing to do with the properties of the individual elements, i.e., sodium and chloride. Looking at NaCl, sodium is known as a very reactive element and chlorine is a poisonous gas, but the table salt has neither of these properties, and almost everyone uses this every day without any bad effect. This indicates that ionic bond is a very strong bond. In case of oppositely charged ions with different charge magnitude, a stoichiometry is maintained so as to establish electrical neutrality. For example, one Mg^{2+} ion is bonded to either one O^{2-} ion or two Cl^- ions forming MgO and $MgCl_2$, respectively.

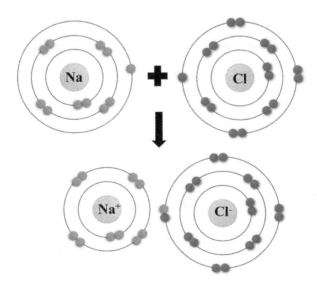

FIGURE 1.1 Formation of NaCl by ionic bonding between Na^+ and Cl^-.

1.2.2 Covalent Bond

In contrast to ionic bonding, in a covalent bonding, some common electrons are basically shared among two atoms. This bond is mostly formed between atoms of the same element or atoms of the elements that are closely placed in the periodic table. To be more specific, this type of bonding is usually feasible for nonmetals and between metals and nonmetals. Electronegativity of both the atoms being almost the same, neither of them are expected to donate electron(s), and at the same time, these elements normally exhibit high ionization energy. Hence, the best possible way for them is to share some of their outermost shell electrons with each other in order to obtain the stable octet configuration. Carbon is one of the most suited elements for covalent bond, as it is having four electrons in its outer shell. Donating or accepting four electrons to achieve the octet configuration is highly energetically unfavorable. Hence, the carbon atom shares its four electrons with four other electrons by making covalent bonds to achieve stability. For example, hydrogen is having only a single electron in its outer shell and thus requires another electron to achieve stable gas configuration. In order to satisfy this, the four outermost electrons of a single carbon atom form covalent bonds with four such hydrogen atoms as shown in Figure 1.2. By doing so, both carbon and hydrogen obtain the stable gas configuration.

Similarly, another common example of covalent bond is formation of hydrogen gas (H_2). Each hydrogen atom shares its electron with another atom to achieve stable gas configuration. The strong attraction force or bonding in case of a covalent bond is due to the electrostatic attraction of the shared pair of electrons to the nuclei of both the atoms. For most of the gases, covalent bond is quite preferable.

1.2.3 Metallic Bond

This type of bond is typically prominent in the case of metals in which there is lack both in the availability of oppositely charged ions and in the number of electrons required for covalent bond. Usually, metals are having a tendency to throw their electrons from valence band to the conduction band owing to their high electrical conductivity. This leads to formation of a negatively charged electron cloud around

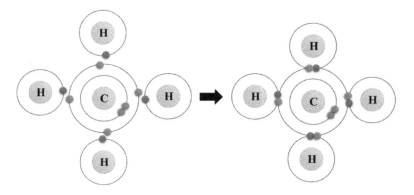

FIGURE 1.2 Formation of methane by covalent bonds between carbon and hydrogen atoms.

An Introduction to Metals

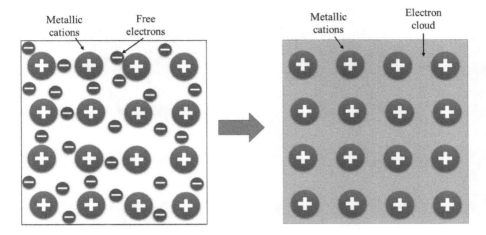

FIGURE 1.3 Illustration of metallic bonding.

the atom. Metallic bond in this case is thus defined as the electrostatic attraction between the positively charged nucleus and the negatively charged electron cloud. In other words, it can also be stated as the sharing of the conduction electrons in a positively charged cation's structure. The metallic atoms thus are bonded together due to their attraction toward the shared sea/cloud/swarm of free electrons around them as shown in Figure 1.3. As this is a kind of sharing of free electrons among the atoms as explained in covalent bonding, metallic bond very often can be considered as extended covalent bond.

Many of the physical and mechanical properties of metals, such as density, strength, ductility, and electrical and thermal conductivity, are due to this metallic bonding among the atoms.

1.2.4 VAN DER WAALS FORCES

van der Waals forces are not of the nature of chemical bonds and usually distance dependent. Unlike other atomic bonds explained earlier, these bonds are relatively weak forces and gradually diminished with increasing distance between participating atoms/molecules. If no other force is present among neighboring atoms, then there exists a critical distance between them after which their nuclei start repelling each other due to similar nature of charge. Very often, van der Waals forces are also termed as intermolecular forces. Dipole–dipole forces, dispersion forces, and hydrogen bonding are different types of van der Waals forces. van der Waals forces are operative at lower temperatures where thermal agitation of atoms is minimum.

1.3 CRYSTAL STRUCTURES

In a crystalline material, atoms occupy some specific points in the space. An example of atomic arrangement is shown in Figure 1.4a. Atomic positions (center of the atom) can be thought of the points of intersection of an array of infinite length lines.

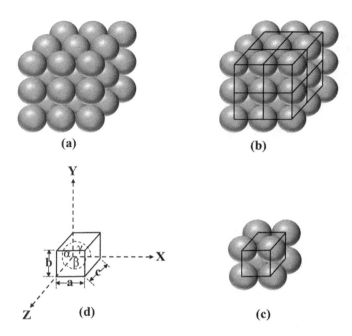

FIGURE 1.4 Crystalline arrangement: (a) an array of atoms in space, (b) construction of space lattice (a network of lines), (c) a representative volume element (unit cell), and (d) parameters of a unit cell.

This array of straight lines is termed as space lattice (Figure 1.4b). These atomic positions are identical to each other, or in other words, all the atoms are having the same surroundings. A basis or motif can be defined as a group of atoms that are located at a particular order with respect to each other. Combination of both the lattice and motif gives rise to the complete crystal structure of the atoms. A unit cell in a crystal is further defined as the smallest representative volume element (Figure 1.4c), which upon repetition in all three dimensions yields the complete crystal structure. A unit cell is a fundamental entity of all the crystalline materials. A unit cell is constituted by its six parameters, three intersecting axial lengths (sides), and correspondingly three angles between each pair of axes. The unit cell dimensions in the X, Y, and Z (not necessarily perpendicular to each other) directions are normally represented by a, b, and c, respectively. α, β, and γ represent the angles between Y and Z, Z and X, and X and Y, respectively, as shown in Figure 1.4d.

All the possible space lattices can be formed from seven different crystal systems or unit cells. These seven crystal systems are different from each other based on the values of a, b, c, α, β, and γ. These possible crystal systems are reported in Table 1.1.

Interestingly, most of the important metals crystallize in two particular crystal systems, i.e., cubic and hexagonal, and only three Bravais lattices are common such as face-centered cubic (FCC), body-centered cubic (BCC), and close-packed hexagonal (CPH).

TABLE 1.1
Various Possible Crystal Systems and Corresponding Bravais Lattices

Crystal System	Sides	Angles	Possible Bravais Lattices
Triclinic	$a \neq b \neq c$	$\alpha \neq \beta \neq \gamma \neq 90°$	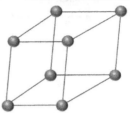 Simple triclinic
Monoclinic	$a \neq b \neq c$	$\alpha = \gamma = 90° \neq \beta$	 Simple monoclinic 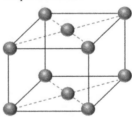 Base-centered monoclinic
Hexagonal	$a = b \neq c$	$\alpha = \beta = 90°, \gamma = 120°$	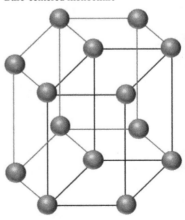 Simple hexagonal

(*Continued*)

TABLE 1.1 (*Continued*)
Various Possible Crystal Systems and Corresponding Bravais Lattices

Crystal System	Sides	Angles	Possible Bravais Lattices
Rhombohedral/ trigonal	$a = b = c$	$\alpha = \beta = \gamma \neq 90°$	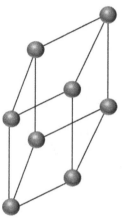 Simple trigonal
Orthorhombic	$a \neq b \neq c$	$\alpha = \beta = \gamma = 90°$	 Simple orthorhombic 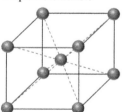 Body-centered orthorhombic 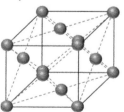 Face-centered orthorhombic

(*Continued*)

An Introduction to Metals

TABLE 1.1 (*Continued*)
Various Possible Crystal Systems and Corresponding Bravais Lattices

Crystal System	Sides	Angles	Possible Bravais Lattices
			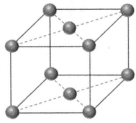 Base-centered orthorhombic
Tetragonal	$a = b \neq c$	$\alpha = \beta = \gamma = 90°$	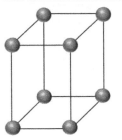 Simple tetragonal
			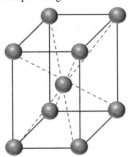 Body-centered tetragonal
Cubic	$a = b = c$	$\alpha = \beta = \gamma = 90°$	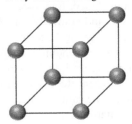 Simple cubic

(*Continued*)

TABLE 1.1 (*Continued*)

Various Possible Crystal Systems and Corresponding Bravais Lattices

Crystal System	Sides	Angles	Possible Bravais Lattices

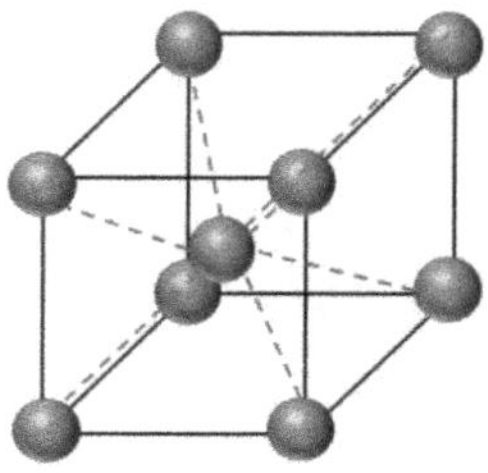

Body-centered cubic

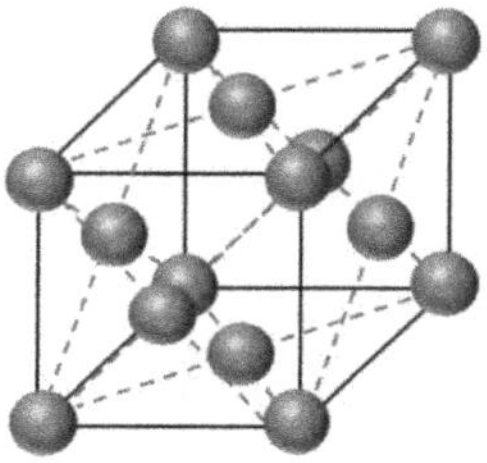

Face-centered cubic

1.3.1 FACE-CENTERED CUBIC STRUCTURE

In this structure, the unit cell is cubic in nature ($a = b = c$ and $\alpha = \beta = \gamma = 90°$), and the atoms occupy the cube corners and face centers as shown in Table 1.1. An infinite array of atoms with FCC arrangement is shown in Figure 1.5a. The smallest repeating unit of this can thus be extracted as shown in Figure 1.5b. Consider the right-hand-side face of the cube in Figure 1.5b. This face is common to two unit cells, one shown in the figure and the other just immediate next right to it.

As can be seen, the corner atoms are not touching each other. The face center atom of this right-side face (shown as atom "X") touches the four corner atoms on that particular face, four face center atoms (top, bottom, front, and back) of the shown unit cell, and similarly four face center atoms of the adjacent right unit cell. Thus, an atom is having 12 nearest neighboring atoms, and therefore, the coordination number of an FCC atom is 12. A representative of the atomic location in the unit cell is shown in Figure 1.5c. However, note that in this figure the spheres (corner and face center) represent the center of the atom (readers should not assume the size of this sphere to be the same as that of the corresponding atom; it is just for the sake of simple representation).

Many of the ductile metals such as copper, aluminum, nickel, silver, and γ-iron crystallize in the FCC form.

An Introduction to Metals

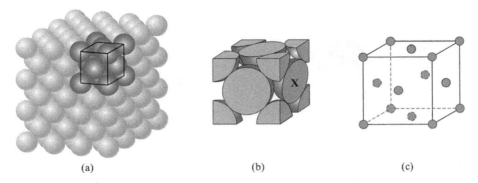

FIGURE 1.5 Face-centered cubic structure: (a) arrangement of atoms, (b) atomic position in a single unit cell, and (c) representative unit cell with atomic locations.

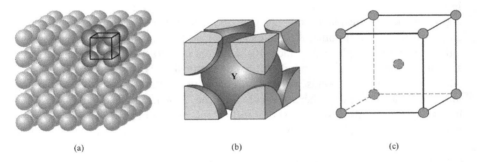

FIGURE 1.6 Body-centered cubic structure: (a) arrangement of atoms, (b) atomic position in a single unit cell, and (c) representative unit cell with atomic locations.

1.3.2 Body-Centered Cubic Structure

In a similar way, a BCC crystal exhibits a cubic unit cell with atoms located at all the cube corners and at the body center. The corner atoms do not touch each other. However, the body-centered atom (shown as atom "Y") touches all the eight corner atoms of the unit cell. Thus, the number of nearest neighbors or coordination number of a BCC atom is eight. The array of atoms, atomic arrangement in a unit cell, and a representative unit cell have been represented in Figure 1.6a, b, and c, respectively.

Metals such as vanadium, sodium, tungsten, barium, and α-iron crystallize in the BCC form.

1.3.3 Close-Packed Hexagonal

A CPH or hexagonal close-packed (HCP) structure usually comprises two parallel planes containing 2-D hexagonal lattices with atoms placed at the corners and center of the hexagons (Figure 1.7a). These closely packed planes are known as basal planes (or A plane). In addition, on the horizontal plane (or B plane), which is located

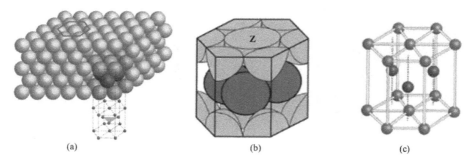

FIGURE 1.7 Close-packed hexagonal structure: (a) arrangement of atoms, (b) atomic position in a single unit cell, and (c) representative unit cell with atomic locations.

exactly at the vertical center of these two basal planes, three more atoms are present in the form of a triangle as shown in Figure 1.7c. The basal plane can be divided into six identical equilateral triangles, and the projections of the centers of the three atoms of the middle plane coincide with the centroid of the alternate equilateral triangles on the basal plane. As one can visualize, the central atom of a hexagonal lattice on the basal plane (top face of the hexagonal lattice shown in Figure 1.7b, shown as atom "Z") makes bond with each of the six hexagonal corner atoms. In addition, it also forms bonds with each three atoms of the B plane enclosed in the unit cell. In a similar way, there also exists another B plane above the top face of the hexagon corresponding to the next unit cell. Three atoms of this B plane form bonds with atom "Z." Thus, the coordination number of a CPH atom is 12. The vertical line joining the face-centered atoms of the top and bottom surface passes through the centroid of the equilateral triangle formed by the three atoms in the middle plane. Metals such as cadmium, cobalt, zinc, magnesium, and α-titanium crystallize in the CPH structure.

If one closely observes the atomic arrangements of FCC and CPH, there is a very little difference. Figure 1.8 is provided to figure out the structural difference in arrangement of atoms in FCC and CPH. Let us first consider a layer of closely packed atoms as shown in Figure 1.8a. This is the most closely packed configuration of atoms, or atomic plane with least possible free space. In this arrangement, each atom is touching and surrounded by six atoms. Let this position be called as A plane, i.e., the centers of all the atoms are placed in "A" position as shown in Figure 1.8b. Now, there are two types of void location in this plane, i.e., "B" and "C." That means for close packing in the z-direction, the centers of the atoms of the next layer must be placed either vertically above "B" or above "C." Let us assume the next layer to be of "B" type as shown in Figure 1.8c. After this, the next or third layer will decide whether it will be of FCC or CPH structure. After placement of the second atomic layer at "B" position, the centers of the atoms of the third layer may be placed either vertically above "A" or above "C." If it is placed at "A" position, it is termed as "ABCABC..." stacking or FCC configuration as shown in Figure 1.8d. On the contrary, if the third layer of atoms are placed at "A" position, it is termed as "ABAB..." stacking, and consequently, the arrangement is known as CPH packing as shown in Figure 1.8e.

An Introduction to Metals

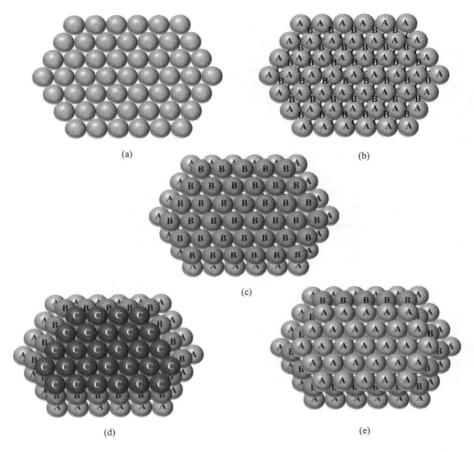

FIGURE 1.8 Layer-by-layer arrangement of atomic planes in the case of FCC and CPH. FCC, face-centered cubic; CPH, close-packed hexagonal.

Hence, the stacking sequence of atomic planes or the position of the atomic centers of the atomic planes makes FCC different from CPH. However, it is very evident from the previous explanation that the atomic packing is the same for both the structures.

1.3.4 Allotropy

The term "Allotropes" comes from the Greek, where "allos" means another and "tropes" means manner or style. Hence, allotropes means that an element may crystallize in different crystal structure in its solid form. The best example of allotropes is iron. When a liquid iron is cooled down at normal atmospheric pressure, it solidifies at 1539°C, has BCC crystal structure, and is known as δ-iron. Upon further cooling, this δ-iron converts to γ-iron at 1394°C. γ-Iron has FCC crystal structure and is stable up to 912°C after which it again converts to a BCC structure known as α-iron. β-Iron also exists but mostly becomes obsolete these days having the same crystal structure

14   Phase Transformations and Heat Treatments of Steels

as α-iron, but having paramagnetic behavior. Upon heating the ferromagnetic α-iron, it loses its magnetism above the Curie temperature, which is 770°C, and becomes paramagnetic β-iron. However, the crystal structures of both α-iron and β-iron are identical. Another well-known element for allotropes is carbon. Carbon crystallizes with several allotropic forms, i.e., diamond having tetrahedral structure, graphite having hexagonal structure, and fullerenes including Bucky balls (C_{60}). Other metals such as tin, cobalt, and polonium also exhibit allotropes.

1.3.5 PACKING FACTOR/EFFICIENCY

Packing factor or packing fraction (PF) or packing efficiency of a crystal structure is defined as the ratio of the volume of atoms enclosed in a unit cell to the total volume of the unit cell. This is basically an entity to estimate the extent of packing or, in other words, the extent of available free space in the crystal structure. Let us thus find out the PF of the aforementioned three most common crystal structures.

As per the definition of PF, it can be mathematically represented as

$$PF = \frac{\text{Total volume of the atoms enclosed in an unit cell}}{\text{Total volume of the unit cell}}$$

The next term now comes is the effective number of atoms. Very often, an atom in a unit cell does not completely belong to that particular atom; rather, it is shared among some adjoining unit cells. For example, a face-centered atom is shared among two unit cells. From this, the PF can be further simplified as

$$PF = \frac{\text{No. of effective atoms} \times \text{volume of a single atom}}{\text{Total volume of the unit cell}}$$

The number of effective atoms is basically the total number of atoms that solely correspond to only a single unit cell. This can be further mathematically defined as

$$\text{No. of effective atoms} = \sum_i C_i n_i$$

C_i = contribution of the ith type of atoms to the unit cell, and
n_i = the total number of ith type of atoms in the unit cell.

In this way, let us find out the PF of the crystal structures.

1.3.5.1 Packing Fraction of a Body-Centered Cubic Structure

Let us consider a BCC single unit cell ABCDEFGH as discussed earlier and shown in Figure 1.9a. Here, there are two types of atoms: (i) cube corner atoms and (ii) body center atoms.

A cube corner is basically common to eight adjacent cubes. Hence, the atom at the cube corner is shared by the same eight unit cells. As a consequence, $\frac{1}{8}$th volume of the atom is enclosed by a single unit cell. Thus, $C_C = \dfrac{1}{8}$ for the cube corner atom

An Introduction to Metals

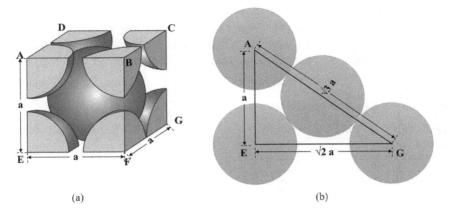

FIGURE 1.9 Atomic arrangement of BCC crystal structure. BCC, body-centered cubic.

in this BCC structure, and there are eight such cube corners making $n_C = 8$. Then, consider the body center atom. The entire body center atom is enclosed in a single unit cell. Thus, $C_{BC} = 1$, and there is only one such body center atom in a unit cell ($n_{BC} = 1$).

$$\text{No. of effective atoms} = \sum_i C_i n_i = C_C \times n_C + C_{BC} \times n_{BC} = \frac{1}{8} \times 8 + 1 \times 1 = 2$$

Assuming atoms to be of spherical shape and taking the atomic radius to be r and lattice parameter to be a,

$$\text{PF} = \frac{\text{No. of effective atoms} \times \text{volume of a single atom}}{\text{Total volume of the unit cell}} = \frac{2 \times \frac{4}{3} \pi r^3}{a^3}$$

Now, we require a relation between r and a in order to find out the exact PF for the structure. For this, consider the right angle triangle AEG as shown in Figure 1.9b. EG corresponds to the face diagonal of the side EFGH having square shape. Thus, the length of EG becomes "$\sqrt{2}a$" (Pythagorean theorem) as shown in the figure. The size and location of the atoms are also clearly visible in the figure (i.e., the body center atom touches all other atoms). Thus, the diagonal AG of the right angle triangle AEG is "$\sqrt{3}a$." From the figure, it can also be seen that AG is four times the radius of the atom, i.e., AG = $4r$. Hence,

$$\sqrt{3}a = 4r$$

$$\text{Hence, PF} = \frac{2 \times \frac{4}{3} \pi r^3}{a^3} = \frac{\frac{8}{3} \pi r^3}{\left(\frac{4}{\sqrt{3}} r\right)^3} = \frac{\frac{8}{3} \pi}{\frac{64}{3\sqrt{3}}} = 0.68$$

Hence, for a BCC structure, the packing efficiency is 0.68 or 68%.

1.3.5.2 Packing Fraction of a Face-Centered Cubic Structure

In a similar way, the PF of an FCC crystal structure can also be determined. Consider the unit cell of FCC crystal structure ABCDEFGH as shown in Figure 1.10a. The atoms are located at the cube corners and at the face centers. Each face center atom is shared by two adjacent unit cells having a common face. Hence, $C_{BC} = \frac{1}{2}$, and there are total such face center atoms in each unit cell, i.e., $n_{BC} = 6$. As explained in the case of BCC, in the FCC structure also, $C_C = \frac{1}{8}$ and $n_C = 8$. The number of effective atoms is thus

$$\text{No. of effective atoms} = \sum_i C_i n_i = C_C \times n_C + C_{FC} \times n_{FC} = \frac{1}{8} \times 8 + \frac{1}{2} \times 6 = 4$$

Assuming atoms to be hard spheres, the packing factor now is

$$PF = \frac{\text{No. of effective atoms} \times \text{volume of a single atom}}{\text{Total volume of the unit cell}} = \frac{4 \times \frac{4}{3}\pi r^3}{a^3}$$

Now, to establish a relation between a and r, consider the right angle triangle AEF as shown in Figure 1.10b, where it can be clearly noticed that the face-centered atom is touching all the corner atoms of the face. Hence, the diagonal AF of the right angle triangle AEF is

$$AF = \sqrt{2}a = 4r$$

Hence,

$$PF = \frac{4 \times \frac{4}{3}\pi r^3}{a^3} = \frac{4 \times \frac{4}{3}\pi r^3}{\left(2\sqrt{2}r\right)^3} = 0.74$$

Therefore, FCC is a denser structure than BCC having a PF of 0.74 or 74%.

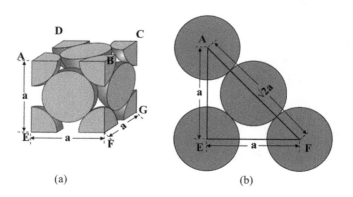

FIGURE 1.10 Atomic arrangement of FCC crystal structure. FCC, face-centered cubic

An Introduction to Metals

1.3.5.3 Packing Fraction of a Close-Packed Hexagonal Structure

For obtaining the PF of a CPH structure, one first needs to determine the volume of the unit cell. Unlike a cubic structure, the sides of a hexagonal structure are not equal, i.e., $a = b \neq c$.

If one clearly observes the arrangement of atoms in the CPH structure, it can be noticed that in a unit cell, there exists a middle horizontal plane containing three atoms, which is exactly midway between the two parallel hexagons (basal planes) as shown in Figure 1.11a. The length of each side of the hexagon and the vertical distance between the hexagons are represented by "a" and "c," respectively. As the basal plane is the most closely packed plane, each atom touches six neighboring atoms. Thus, the side of the hexagon is equal to the diameter of the atoms ($a = 2r$, r is the radius of the atom). Furthermore, each of the hexagons of the unit cell can be divided into six equilateral triangle as shown in Figure 1.11b. And the centers of the atoms in the middle plane exist exactly at the middle points of the vertical lines joining the centroids of each pair of the alternate equilateral triangles as shown in Figure 1.11b. A simplified version of this arrangement has been presented in Figure 1.11c. The atom "D" (an atom of the middle plane) sits just vertically above the centroid of the equilateral triangle formed by A, B, and C atoms. However, all these four atoms

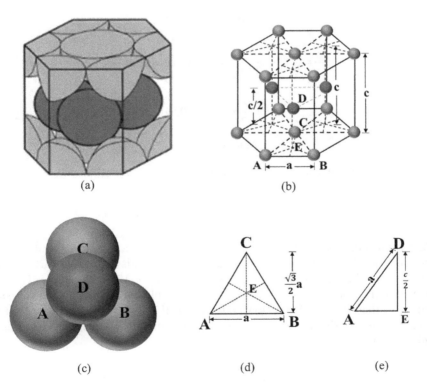

FIGURE 1.11 Arrangement of atoms in CPH structure and determination of its PF. CPH, close-packed hexagonal; PF, packing fraction.

(A, B, C, and D) touch each other (discussed earlier). The volume of the hexagonal unit cell can be obtained by multiplying the area of the hexagon with the height. Therefore, a relation must be established between the lattice parameters "c" and "a" in order to represent the volume only in terms of "a." For that, consider the equilateral triangle obtained by the centers of atoms A, B, and C as shown in Figure 1.11d. E is the centroid of this triangle, and hence, the center of atom D is exactly vertically above the point E. As the distance between the vertex and centroid of the triangle is two-thirds of the length of the median,

$$AE = \frac{2}{3} \times \sqrt{a^2 - \frac{a^2}{4}} = \frac{2}{3} \times \frac{\sqrt{3}}{2} a = \frac{a}{\sqrt{3}}$$

Now, again a right angle triangle can be visualized, which is obtained by the points A, E, and D as shown in Figure 1.11e. As the atom D touches all the atoms A, B, and C, the distance between the centers of atoms D and A is equal to the diameter of the atoms, which is again equal to the lattice parameter "a." The vertical distance between the centers of the atoms A and D is half of the height of the hexagon, i.e.,

$$DE = \frac{c}{2}$$

$$As, DE = \sqrt{AD^2 - AE^2}$$

$$\frac{c}{2} = \sqrt{a^2 - \frac{a^2}{3}} \quad \left(As\ AD = a\ and\ AE = \frac{a}{\sqrt{3}} \right)$$

$$Or, \frac{c}{2} = \sqrt{\frac{2}{3}} a = 0.816a$$

$$Or,\ c = 1.632a$$

Now, Total volume of the unit cell = area of the hexagon × height of the unit cell

$$Total\ volume\ of\ the\ unit\ cell = \frac{3\sqrt{3}}{2} a^2 \times c = \frac{3\sqrt{3}}{2} a^2 \times 1.632a = 4.24\,a^3 = 33.92\,r^3$$

Each atom at the corner of the hexagon is shared by six such unit cells, and thus, the contribution of a single corner atom to a single unit cell is $\frac{1}{6}$, and there are six such corner atoms in each of the hexagon making a total of 12 corner atoms. The face center atom at the center of each basal hexagonal plane is shared among two such unit cells; thus, its contribution toward the unit cell is $\frac{1}{2}$, and there are two such face center atoms in a single unit cell. All the three atoms in the middle plane belong to only a single unit cell. Thus,

$$No.\ of\ effective\ atoms = \frac{1}{6} \times 12 + \frac{1}{2} \times 2 + 1 \times 3 = 6$$

An Introduction to Metals 19

Therefore,

$$PF = \frac{No.\,of\,effective\,atoms \times volume\,of\,a\,single\,atom}{Total\,volume\,of\,the\,unit\,cell} = \frac{6 \times \frac{4}{3}\pi r^3}{33.92\,r^3}$$

Thus, $PF = 0.74$ or 74%

From the previous discussion, it is very clear that both FCC and CPH have the same PF, which is also quite expected due to the layer-by-layer stacking of closely packed atomic planes as discussed earlier. However, the BCC structure is less densely packed in comparison to FCC and CPH.

FURTHER READING

Avner, S. H. *Introduction to Physical Metallurgy*. (Tata McGraw-Hill Education, New York, 1997).

Azaroff, L. V. *Introduction to Solids*. (Tata McGraw-Hill Education, New York, 1960).

2 Diffusion

2.1 ATOMIC DIFFUSION MECHANISMS

Atomic diffusion in solids only occurs when atoms jump from one site to another. Normally, atoms in a crystal oscillate around their equilibrium positions, and when these oscillations become large enough, atoms change sites. The nature of atomic movement in the solid decides the diffusion mechanism, i.e., interstitial and vacancy mechanism.

Interstitial mechanism – an atom is said to diffuse by an interstitial mechanism when it passes from one interstitial site to one of its nearest-neighbor interstitial sites without permanently displacing any of the matrix atoms. For example, let us consider the interstitial diffusion mechanism in the (100) plane of an FCC lattice, as shown in Figure 2.1. The atom labeled 3 represents the position of the interstitial atom before diffusion whereas the atom labeled 4 represents the position after diffusion. However, the jump is only possible if the matrix atoms labeled 1 and 2 move apart to allow the atom labeled 3 to go through them and occupy the position of the atom labeled 4. The movement of matrix atoms will cause distortion, and it is this distortion that constitutes the barrier to an interstitial atom changing sites. Generally, the interstitial diffusion mechanism occurs in alloys, e.g., C in α- and γ-irons. It also occurs in substitutional alloys like in radiation damage studies, energetic particles, e.g., neutrons, can knock atoms off normal lattice sites into interstitial positions to form "self-interstitials".

Vacancy mechanism – the atom is said to have diffused by a vacancy mechanism if one of the atoms on an adjacent site jumps into the vacancy. Let us consider a close-packed plane of spheres in an FCC lattice, as shown in Figure 2.2. The atom

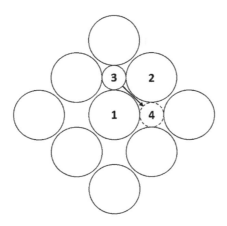

FIGURE 2.1 A schematic representation of interstitial diffusion mechanism on a FCC (100) plane. FCC, face-centered cubic.

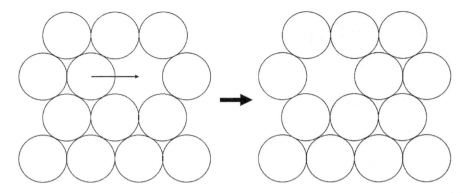

FIGURE 2.2 Movement of an atom by vacancy diffusion in an FCC lattice. FCC, face-centered cubic.

diffuses to the vacant site, creating a vacancy on the site that the atom has just left. Hence, the diffusion will depend upon the presence of a vacancy on an adjacent site, and the rate of diffusion is dependent upon how easily vacancies can form in the lattice and how easy it is for an atom to move into a vacancy.

2.2 TYPES OF DIFFUSION

2.2.1 INTERSTITIAL DIFFUSION

There are two common types of diffusion process. *Substitutional* diffusion occurs when substitutional atoms diffuse by the vacancy mechanism. On the other hand, in the case of interstitial atoms, diffusion takes place by forcible movement of the interstitial atom through the parent larger atoms, known as *interstitial* diffusion. The interstitial diffusion process is discussed in detail in the following.

Consider a simple cubic arrangement of atoms and the solute atoms occupying the interstitial positions. The atomic planes 1 and 2 contain $n1$ and $n2$ atoms per square meter, respectively. Consider a very dilute solution such that each solute atom is having six neighboring interstitial sites, which are unoccupied. In such case, if each interstitial atom jumps γ times per second, then the number of atoms that will jump from plane 1 to 2 in 1 s is given by

$$J_{1-2} = \frac{1}{6}\gamma n_1 \text{ atoms/m}^2/\text{s} \tag{2.1}$$

In the same way, the number of atoms that will jump from plane 2 to 1 is given by

$$J_{2-1} = \frac{1}{6}\gamma n_2 \text{ atoms/m}^2/\text{s} \tag{2.2}$$

As $n_1 > n_2$, a net atom flux will be evolved from plane 1 to 2, which is represented by

$$J_{\text{net}} = J_{2-1} - J_{1-2} = \frac{1}{6}\gamma(n_2 - n_1) \text{ atoms/m}^2/\text{s} \tag{2.3}$$

Diffusion

If the jump distance or the separation between planes 1 and 2 is α, then the relation between n_1 and the concentration of solute atom at the position of plane 1, C_1 will be

$$C_1 = \frac{n_1}{\alpha} \tag{2.4}$$

Similarly,

$$C_2 = \frac{n_2}{\alpha} \tag{2.5}$$

Hence, from equations 2.4 and 2.5, we have

$$n_1 - n_2 = \alpha(C_1 - C_2) \tag{2.6}$$

But, from Figure 2.3, we have

$$C_1 - C_2 = -\alpha \frac{\partial C}{\partial x} \tag{2.7}$$

Therefore, substituting equations 2.6 and 2.7 in equation 2.3 will give

$$J_{net} = -\left(\frac{1}{6}\gamma\alpha^2\right)\frac{\partial C}{\partial x} \text{ atoms}/m^2/s \tag{2.8}$$

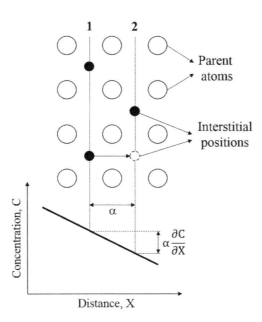

FIGURE 2.3 Interstitial diffusion process in a concentration gradient (the derivation of Fick's first law).

The concentration gradient in the above case $\partial C/\partial x$ is a partial derivative and can be a function of time. Hence, it is obvious that there will be a flow of atoms in the downward direction of the concentration gradient by a random jumping way.

Replacing $D = \dfrac{1}{6}\gamma\alpha^2$ in equation 2.8 gives

$$J_{net} = -D\dfrac{\partial C}{\partial x} \qquad (2.9)$$

This equation is popularly known as Fick's first law of diffusion, and D represents the diffusion coefficient or intrinsic diffusivity. D has the unit of m²/s. Steady-state diffusion is described by Fick's first law which states that the concentration does not change with time. Flux, which is the net number of atoms crossing a unit area perpendicular to a given direction per unit time, is constant with time. However, for most practical situations, concentration varies with distance and time. Such a diffusion is called non–steady-state diffusion. In such cases, Fick's first law can no longer be valid, and therefore, we need to find out the variation of concentration with distance and time. Let us consider a simple case, where there exists a concentration profile in a single direction (x) only (Figure 2.4).

The number of interstitial atoms that diffuse into the thin slice of area A and thickness δx along the plane 1 in a very small time interval δt may be represented by $J_1 A \delta t$. Similarly, the number of atoms that leave the slice across the plane 2 during this time δt is $J_2 A \delta t$. Since some amounts of interstitial atoms get accumulated in the thin slice, $J_2 < J_1$. So, the concentration of interstitial atoms within the thin slice is given by

$$\delta C = \dfrac{(J_1 - J_2)A\delta t}{A\delta x} \qquad (2.10)$$

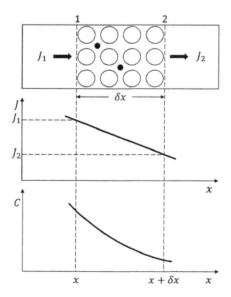

FIGURE 2.4 The derivation of Fick's second law.

Diffusion   **25**

But δx is small. So, we have

$$J_2 = J_1 + \frac{\partial J}{\partial x}\delta x \qquad (2.11)$$

$$\Rightarrow J_1 - J_2 = -\frac{\partial J}{\partial x}\delta x \qquad (2.12)$$

Substituting the value of $J_1 - J_2$ in equation 2.10, we get

$$\delta C = \left(-\frac{\partial J}{\partial x}\right)\delta t \qquad (2.13)$$

As $\delta t \to 0$, we have

$$\frac{\partial C}{\partial t} = -\frac{\partial J}{\partial x} \qquad (2.14)$$

From Fick's first law, we know that

$$J = -D\frac{\partial C}{\partial x} \qquad (2.15)$$

Combining equations 2.14 and 2.15, we get

$$\frac{\partial C}{\partial t} = \frac{\partial}{\partial x}\left(D\frac{\partial C}{\partial x}\right) \qquad (2.16)$$

This expression is commonly called as Fick's second law of diffusion. In some cases, D can be assumed to be independent of concentration, and under such condition, equation (2.16) can further be simplified as

$$\frac{\partial C}{\partial t} = D\frac{\partial^2 C}{\partial x^2} \qquad (2.17)$$

A graphical representation of equation is presented in Figure 2.5. Figure 2.5a shows that the concentration at all points increases with time, i.e., $\frac{\partial^2 C}{\partial x^2} > 0$, whereas Figure 2.5b shows that the concentration at all points decreases with time, i.e., $\frac{\partial^2 C}{\partial x^2} < 0$.

2.2.1.1 Solution to the Diffusion Equation

The solution will be discussed in case of carburization of steel, which is commonly encountered and has significant practical importance. During carburization of steel, the carbon content at the surface is maintained at some constant value by controlling the amounts of gas mixtures in which the steel is held for a certain period of time.

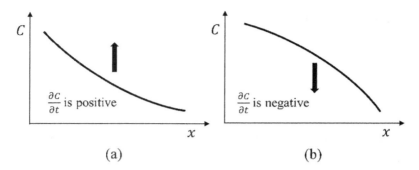

FIGURE 2.5 Concentration profile with time: (a) positive curvature and (b) negative curvature.

Simultaneously, diffusion of carbon also takes place from the surface to the core of the steel. The concentration profiles along distance were then plotted at different time periods as shown in Figure 2.6. A more generalized mathematical formulation then can be developed by taking into account the solved Fick's second diffusion law with some boundary conditions as stated in the following:

$$C(\text{at } x = 0) = C_S \tag{2.18}$$

$$C(\infty) = C_0 \tag{2.19}$$

where C_0 is the original carbon concentration of the steel.

Consider the case where the diffusivity of carbon in austenite has a direct relation with the concentration, and taking an infinite long specimen, an average value may be helpful in order to attain the solution. In this case, the solution can be written as

$$C = C_S - (C_S - C_0)\operatorname{erf}\left(\frac{x}{2\sqrt{Dt}}\right) \tag{2.20}$$

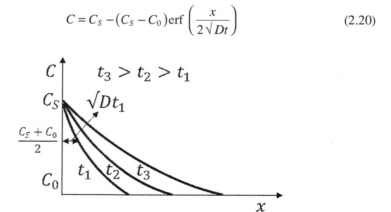

FIGURE 2.6 Concentration profiles at different time intervals. C_0 is the initial concentration of the steel, and C_S is the surface concentration, which is maintained constant.

Diffusion

where "erf" is the usual notation for error function. The mathematical expression for "erf" can be defined as

$$\text{erf}(z) = \frac{2}{\sqrt{\pi}} \int_0^z \exp(-y^2) \, dy \qquad (2.21)$$

The error function values can be found from books of standard mathematical functions. Since $\text{erf}(0.5) \approx 0.5$, the depth of the carbon concentration is midway between C_S and C_0, which is given by

$$\frac{x}{2\sqrt{Dt}} \approx 0.5 \qquad (2.22)$$

$$\Rightarrow x \approx \sqrt{Dt} \qquad (2.23)$$

From this expression, it can clearly be visualized that there exists a direct relationship between the depth of isoconcentration and $\sqrt{Dt}$. For example, for a diffusivity of carbon in austenite at 1000°C, $D \approx 4 \times 10^{-11}$ m^2/s, a carburized layer of 0.2 mm depth will take a time duration of 1000 s, i.e., $(0.2 \times 10^{-3})^2 / 4 \times 10^{-11}$. Similarly, there exist many other examples apart from carburization of steel where the solution of the diffusion equation can be applied. A summary of the solutions to different practical situations is described in Table 2.1.

2.2.1.2 Effect of Temperature

Temperature has a vital influence on the diffusivity and diffusion rates. To find out the effect of temperature on the interstitial diffusion in solids, let us consider the jump technique of an interstitial atom, as shown in Figure 2.7.

TABLE 2.1

Solutions of the Diffusion Equations Encountered in Different Practical Situations

Carburization	$C = C_S - (C_S - C_0)\text{erf}\left(\dfrac{x}{2\sqrt{Dt}}\right)$
Decarburization	$C = C_0\text{erf}\left(\dfrac{x}{2\sqrt{Dt}}\right)$
Diffusion couple	$C = \left(\dfrac{C_1 + C_2}{2}\right) - \left(\dfrac{C_1 - C_2}{2}\right)\text{erf}\left(\dfrac{x}{2\sqrt{Dt}}\right)$
	C_1 = concentration of steel 1
	C_2 = concentration of steel 2
Homogenization	$C = C_{\text{mean}} + \beta_0 \sin\left(\dfrac{\pi x}{\lambda}\right)\text{erf}\left(-\dfrac{t}{\tau}\right)$
	β_0 = initial concentration amplitude
	λ = half wavelength
	τ = relaxation time

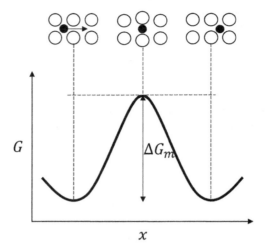

FIGURE 2.7 Effect of position of interstitial atom on the free energy curve.

The atom from the ground state jumps to another ground state through an activated state, where it has to move its neighboring atoms elastically. When the interstitial atom moves through the activated state, the parent atoms are forced apart into higher energy positions. As a result, the net free energy increment for the system is given by ΔG_m (m stands for migration). If the interstitial atom at the initial ground state has a mean frequency of vibration v along the x direction, then it attempts v jumps per second. Out of this, $\exp\left(-\dfrac{\Delta G_m}{RT}\right)$ attempts become eventually successful.

If we consider a 3-D geometry of vibration, each atom, let us say, is surrounded by z identical sites, where accommodation of this atom is possible, and the frequency of this atomic jumping may be mathematically determined by the following expression.

$$\gamma = zv \exp\left(-\frac{\Delta G_m}{RT}\right) \tag{2.24}$$

But we know that $\Delta G_m = \Delta H_m - T\Delta S_m$ and $D = \dfrac{1}{6}\gamma d^2$. So,

$$D = \frac{1}{6} zv \exp\left(-\frac{\Delta G_m}{RT}\right) d^2 \tag{2.25}$$

$$\Rightarrow D = \frac{1}{6} d^2 zv \exp\left(\frac{\Delta S_m}{R}\right) \exp\left(-\frac{\Delta H_m}{RT}\right) \tag{2.26}$$

ΔS_m is due to thermal entropy contribution and independent of temperature. So,

$$D = D_0 \exp\left(\frac{Q_{ID}}{RT}\right) \tag{2.27}$$

Diffusion

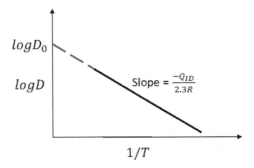

FIGURE 2.8 Plot of plot $\log D$ vs. $1/T$.

where $D_0 = \dfrac{1}{6} d^2 z \nu \exp\left(\dfrac{\Delta S_m}{R}\right)$, and $Q_{ID} = \Delta H_m$ is called the activation enthalpy or interstitial diffusion. This activation energy can be determined from the diffusion coefficients calculated at different temperatures.

$$D = D_0 \exp\left(\dfrac{Q_{ID}}{RT}\right) \tag{2.28}$$

$$\Rightarrow \log D = \log D_0 - \dfrac{Q_{ID}}{2.3RT} \tag{2.29}$$

So if we plot $\log D$ vs. $1/T$, we can determine the activation energy for diffusion Q_{ID} from the slope which is equal to $-\dfrac{Q_{ID}}{R}$ (Figure 2.8).

2.2.2 Substitutional Diffusion

In the case of substitutional diffusion, atom jumps from its position to a nearby vacant lattice position. Hence, a free lattice site in the neighboring sites is the essential criterion for such diffusion to take place. The simplest form of substitutional diffusion is the self-diffusion of atoms in a pure metal.

2.2.2.1 Self-Diffusion

The rate at which self-diffusion in metals takes place may be experimentally determined by incorporating some radioactive atoms (A*) to a pure substance (A). And then the penetration rate may be elucidated at different temperatures. In terms of chemical nature, both A* and A can be treated as identical. Hence, it is fair to assume their jump frequency to be more or less same. Therefore,

$$D_A^* = D_A = \dfrac{1}{6}\gamma d^2 \tag{2.30}$$

In substitutional diffusion, once the atom jumps into the vacancy, the atoms are also having tendency to return back to the earlier location. Under such conditions, these jumps do not contribute to the diffusive flux. So,

$$D_A^* = fD_A = \frac{1}{6}\gamma d^2 \qquad (2.31)$$

where f is the correlation factor. However, the effect of this parameter is negligible, as f is having a value close to unity.

In the case of interstitial diffusion, the probability of successful jumps was given by

$$\exp\left(\frac{-\Delta G_m}{RT}\right)$$

But the adjacent site may not be vacant for substitutional diffusion most of the time. Therefore, the probability for having a vacant neighboring site is zX_v, where z is the number of nearest neighbors. The probability of having any vacant site is referred as X_v, i.e., vacancy mole fraction. Hence, the probability of a successful jump in case of substitutional diffusion is given by

$$zX_v \exp\left(\frac{-\Delta G_m}{RT}\right)$$

Now, the number of such successful jump attempts per unit time performed by an atom, which is vibrating with frequency v, is given by

$$\gamma = vzX_v \exp\left(\frac{-\Delta G_m}{RT}\right) \qquad (2.32)$$

As discussed in earlier chapter, vacancies are thermodynamically stable imperfections, so

$$X_v = X_v^{eq} = \exp\left(\frac{-\Delta G_v}{RT}\right) \qquad (2.33)$$

Substituting equations 2.32 and 2.33 in equation 2.30, we have

$$D_A = \frac{1}{6}d^2 z \exp\frac{-(\Delta G_m + \Delta G_v)}{RT} \qquad (2.34)$$

Equation 2.34 is identical to that of interstitial diffusion with an additional term representing the activation energy required for self-diffusion, i.e., ΔG_v. With this, let us now find out the diffusion coefficient in a body-centered cubic lattice (assuming (110) plane), as shown in Figure 2.9. It is considered that atom 9 lies on the plane of the BCC lattice. The atom can jump to either 5 or 6, which are present on adjacent plane. So, the probability of any atom on (110) plane of a BCC lattice to jump to its adjacent plane is $2/8 = 1/4$, where 8 is the coordination number of the atom. The flux of atoms on plane is, therefore,

$$J_1 = \frac{1}{4}\gamma n_1 \qquad (2.35)$$

Diffusion

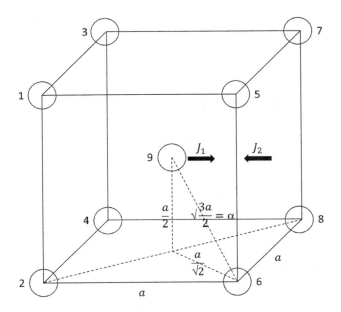

FIGURE 2.9 Self-diffusion process in a (110) plane of a BCC lattice. BCC, body-centered cubic.

where γ is the jump frequency and n_1 is the number of atoms on plane 1. Similarly, the flux of atoms on plane 2 is

$$J_2 = \frac{1}{4}\gamma n_2 \tag{2.36}$$

where n_2 is the number of atoms on plane 2. Since, $n_1 > n_2$,

$$J_{net} = J = J_1 - J_2 = \frac{1}{4}\gamma(n_1 - n_2) \tag{2.37}$$

We know that if C_1 and C_2 are the concentrations of atoms on planes 1 and 2, respectively, then

$$C_1 = \frac{n_1}{d}; C_2 = \frac{n_2}{d} \tag{2.38}$$

From Figure 2.9, we get

$$C_1 - C_2 = -d\frac{\partial C}{\partial x} \tag{2.39}$$

Therefore, we have

$$n_1 - n_2 = -d^2\frac{\partial C}{\partial x} \tag{2.40}$$

32 Phase Transformations and Heat Treatments of Steels

But, d is given by

$$d = \frac{a}{\sqrt{h^2 + k^2 + l^2}} \tag{2.41}$$

where a is the lattice parameter. So,

$$d^2 = \frac{a^2}{2} \tag{2.42}$$

Now, a is related to the jump distance α by

$$\alpha = \frac{\sqrt{3}a}{2} \tag{2.43}$$

Putting the value of a in terms of α in equation 2.42, we have

$$d^2 = \frac{2\alpha^2}{3} \tag{2.44}$$

Hence, equation 2.40 now becomes

$$n_1 - n_2 = -d^2 \frac{\partial C}{\partial x} = -\frac{2}{3} \alpha^2 \frac{\partial C}{\partial x} \tag{2.45}$$

So,

$$J = \frac{1}{4} \gamma \left\{ \frac{2}{3} \alpha^2 \left(-\frac{\partial C}{\partial x} \right) \right\} = \frac{1}{6} \gamma \alpha^2 \left(-\frac{\partial C}{\partial x} \right) = -D \frac{\partial C}{\partial x} \tag{2.46}$$

where $D = \frac{1}{6} \gamma \alpha^2$ is the diffusion coefficient.

2.2.2.2 Diffusion in Substitutional Alloys

In substitutional alloys, there is a difference in the rate at which solute atoms and solvent atoms jump into the nearby available vacant sites. So, each of these atoms must be having its own distinct intrinsic coefficient of diffusion D_A or D_B. According to Fick's first law, we have

$$J_A = -D_A \frac{\partial C_A}{\partial x} \tag{2.47}$$

$$J_B = -D_B \frac{\partial C_B}{\partial x} \tag{2.48}$$

where J_A and J_B represent the fluxes corresponding to the atoms of A and B across a defined crystallographic plane.

Diffusion

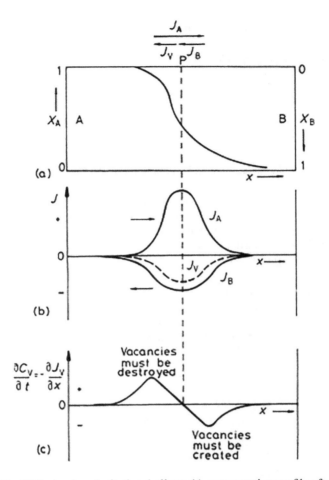

FIGURE 2.10 Diffusion in substitutional alloys: (a) concentration profile after interdiffusion of A and B, (b) variation of J_A, J_B, and J_V with respect to position x, and (c) variation in the formation and destruction rates of vacancies or destroyed across the diffusion couple.

Consider a welded joint made up of pure metals A and B. In such case, interdiffusion takes place among A and B across the welded zone. By allowing diffusion to take place at relatively elevated temperature, a concentration gradient is expected to develop, as represented in Figure 2.10a.

Assumption 1: There is no change in the net number of atoms existing per unit volume of the metal. So,

$$C_O = C_A + C_B \tag{2.49}$$

$$\frac{\partial C_A}{\partial x} = -\frac{\partial C_B}{\partial x} \tag{2.50}$$

This indicates that for both A and B, the driving force for diffusion, i.e., the existing concentration gradient, remains the same in magnitude. However, their gradients are directed in opposite directions. Now,

$$J_A = -D_A \frac{\partial C_A}{\partial x} \quad (2.51)$$

$$J_B = D_B \frac{\partial C_A}{\partial x} \quad (2.52)$$

Assumption 2: $D_A > D_B$. Hence, we have

$$|J_A| > |J_B|$$

Assumption 3: When an atom jumps to a nearby vacancy, it can also be assumed that the earlier vacancy has moved to the earlier location of atom as shown in Figure 2.11.

So, if there exists a net flux of atoms in a particular direction, then it is evident that there must be a net flux of vacancies in the opposite direction. Now, the total flux of vacancies is $-J_A - J_B$, because of the atomic migration of $A(-J_A)$ in addition to the vacancy flux arising due to the diffusion B atoms $(-J_B)$. Since $|J_A| > |J_B|$, net flux of vacancies is

$$J_V = -J_A - J_B \quad (2.53)$$

$$J_V = (D_A - D_B)\frac{\partial C_A}{\partial x} \quad (2.54)$$

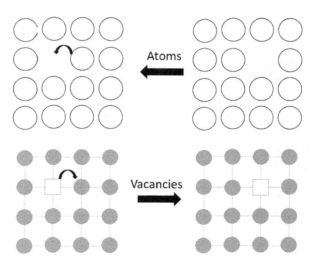

FIGURE 2.11 Schematic representation of mutual exchange of locations among an atom and a vacancy.

Diffusion   **35**

Now, there will be variation in J_V across the diffusion couple, as shown in Figure 2.10b. In order to maintain the vacancy concentration near equilibrium, vacancies must be created on the B-rich side and destroyed on the A-rich side. The rate at which the vacancies are created or destroyed is given by

$$\frac{\partial C_V}{\partial t} = -\frac{\partial J_V}{\partial t} \tag{2.55}$$

This rate of vacancy creation and destruction varies as a function of distance across the couple as represented in Figure 2.10c. Hence, there exists a net flux of vacancies across the central plane of the diffusion couple. It results in the movement of the lattice planes present in the nearby vicinity of the central plane of the couple to the left. The flux of vacancies crossing a lattice plane hence affects the velocity of movement of this lattice plane. Let us suppose that the lattice plane has an area S. So, it will sweep out a volume of $Sv\delta t$ during a small time interval δt. This volume will contain $Sv\delta t C_O$ atoms. Exactly this amount of atoms are displaced by an equal number of vacancies passing through the plane at that time duration, i.e., $J_V S\delta t$. So, we have

$$J_V S\delta t = Sv\delta t C_O \tag{2.56}$$

$$J_V = vC_O \tag{2.57}$$

Combining equations 2.54 and 2.57, we have

$$v = \frac{1}{C_O}(D_A - D_B)\frac{\partial C_A}{\partial x} \tag{2.58}$$

But, we know that $X_A = \dfrac{C_A}{C_O}$. Hence, substituting the value of C_A in terms of X_A gives

$$v = (D_A - D_B)\frac{\partial X_A}{\partial x} \tag{2.59}$$

However, in reality, compositional change may take place at a defined location relative to the specimen ends.

Therefore, there is a need of deriving Fick's second law of diffusion for substitutional alloys. Let us consider a thin slice of material having thickness δx present at a fixed distance x from any of the ends of the diffusion couple. Let J'_A be the total atomic flux through the constant plane having two contributions: one is diffusive flux, $J_V = -D_A\dfrac{\partial C_A}{\partial x}$ and the other is the flux due to the velocity of the lattice in which diffusion is occurring, vC_A. So, we have

$$J'_A = -D_A\frac{\partial C_A}{\partial x} + vC_A \tag{2.60}$$

36 Phase Transformations and Heat Treatments of Steels

Substituting the value of v in equation (2.60), we have

$$J'_A = -D_A \frac{\partial C_A}{\partial x} + C_A (D_A - D_B) \frac{\partial X_A}{\partial x} \tag{2.61}$$

Since, $X_A = \dfrac{C_A}{C_O}$ and $X_B = \dfrac{C_B}{C_O}$, we get

$$J'_A = (-D_A + X_A D_A - X_A D_B) \frac{\partial C_A}{\partial x} \tag{2.62}$$

Simplifying equation (2.62), we have

$$J'_A = -\bar{D} \frac{\partial C_A}{\partial x} \tag{2.63}$$

where $\bar{D}$ is the interdiffusion coefficient, which is defined as follows:

$$\bar{D} = X_A D_B + X_B D_A \tag{2.64}$$

Earlier, we had derived the following relation (see equation 2.14):

$$\frac{\partial C_A}{\partial t} = -\frac{\partial J'_A}{\partial x} \tag{2.65}$$

Hence, substituting equation 2.65 into equation 2.63, we have

$$\frac{\partial C_A}{\partial t} = \frac{\partial}{\partial x} \left(\bar{D} \frac{\partial C_A}{\partial x} \right) \tag{2.66}$$

For substitutional alloys, equation (2.66) is popularly known as Fick's second law for diffusion.

FURTHER READING

Porter, D. A. & Easterling, K. E. *Phase Transformations in Metals and Alloys*, Third Edition (Revised Reprint). (CRC Press, Boca Raton, FL, 1992).

3 Defects in Crystalline Solids

3.1 INTRODUCTION

Any discrepancy from ideal crystalline arrangement in a crystalline solid is known as crystalline defect/imperfection. Imperfections are invariably associated with crystalline solids. The dimension of this defect might be microscopic or macroscopic. Defect might influence most of the properties of the solid, e.g., the type and density of the defects affects the strength, ductility, electrical conductivity, corrosion resistance, etc. Not only this, defect also influences the thermodynamics and kinetics of phase transformations in solids, Hence, it is very important to identify the defect and the possible ways to control them to have a control over the overall properties and performance of the material.

3.2 CLASSIFICATION

Classification of the defects has been made based upon the dimensionality of the defect. The various classes of defects are as follows:

- Zero dimensional (or point defect)
- One dimensional (or line defect)
- Two dimensional (or interfacial defect)
- Three dimensional (or pores, voids, cavities, etc.)

A schematic representation to distinguish these defects from each other is given in Figure 3.1. We shall discuss these defects in more detail in this chapter.

3.2.1 ZERO-DIMENSIONAL OR POINT DEFECT

This defect has a size range of one or two atomic sizes of the same crystal.

3.2.1.1 Vacancy

Vacancy is the simplest point defect in case of metal. When any of the lattice point of a crystalline solid is vacant, a vacancy is said to be created. A pictorial representation of vacancy is shown in Figure 3.1. Among all of the existing crystalline defects, vacancy is the only equilibrium or thermodynamic defect (i.e., creation of vacancy leads to decrease in free energy of the material). The decrease in free energy is associated with the increase in conformational entropy.

To create a vacancy, one of the atoms needs to be removed from its original lattice position. Hence, certain amount of energy is required for this. Let "E_v" be the energy

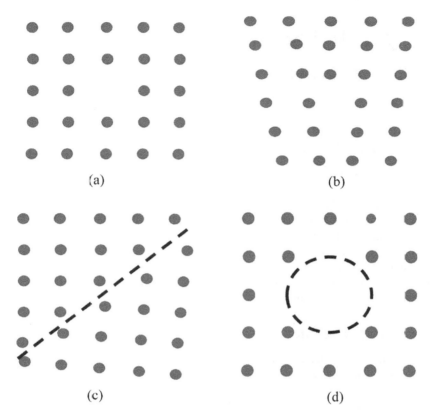

FIGURE 3.1 Schematic representation of crystalline defects: (a) 0-D defect (vacancy), (b) 1-D defect (dislocation), (c) 2-D defect (grain boundary), and (d) 3-D defect (void).

required for creating a single vacancy. Hence, the total increase in energy of the system for n vacancies corresponds to "nE_v" (enthalpic contribution toward free energy).

Next, consider after creation of "n" vacancies, there are "$N-n$" atoms occupying their original lattice positions. Hence, "n" vacancies and "$N-n$" atoms can be arranged over "N" lattice positions by "Ω" ways, where $\Omega = \dfrac{N!}{n!(N-n)!}$.

Hence, the increase in conformational entropy (ΔS) can be calculated as

$$\Delta S = k \ln \Omega \qquad (3.1)$$

Now, the net change in free energy can be determined as

$$\Delta G = nE_v - T\Delta S = nE_v - kT \ln \dfrac{N!}{n!(N-n)!} \qquad (3.2)$$

Defects in Crystalline Solids

Incorporating Stirling's theorem ($\ln a! = a \ln a$),

$$\Delta G = nE_v - kT[N \ln N - n \ln n - (N-n)\ln(N-n)] \quad (3.3)$$

To find out the equilibrium vacancy amount (n_e), the differentiation of "ΔG" with respect to "n_e" should be zero.

$$\frac{\partial(\Delta G)}{\partial n_e} = 0 = E_v - kT[\ln(N - n_e) - \ln n_e]$$

$$E_v = kT \ln \frac{N - n_e}{n_e}$$

$$\ln \frac{N - n_e}{n_e} = \frac{E_v}{kT}$$

$$\ln \frac{n_e}{N - n_e} = -\frac{E_v}{kT}$$

$$\frac{n_e}{N - n_e} = \exp\left[-\frac{E_v}{kT}\right] \quad (3.4)$$

Usually, the relative number of vacancy (n_e) is very less in comparison to the number of lattice positions (N) in any crystal, so "$N - n_e$" can be considered as "N". Hence,

$$\frac{n_e}{N} = \exp\left[-\frac{E_v}{kT}\right] \quad (3.5)$$

This is called the equilibrium vacancy concentration.

A schematic of the variation of free energy with respect to vacancy concentration is shown in Figure 3.2. Vacancy may be created during solidification of the crystal due to hindrance to crystal growth from neighboring crystal nuclei. Atomic diffusion

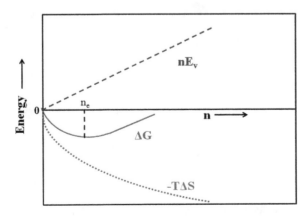

FIGURE 3.2 Variation of free energy with vacancy concentration.

may also be responsible for creation of vacancy in solid state itself. In solid state also, heavy plastic deformation, rapid quenching, high energy particle bombardment, and so forth are some ways of incorporating more vacancies to the system. However, these nonequilibrium vacancies have a tendency to cluster, resulting formation of divacancy or trivacancy and so on. Position exchange of a vacancy with its neighboring atom is an important phenomenon, which is known as diffusion in solids, and is accelerated upon excursion to elevated temperature.

3.2.1.2 Self-Interstitial

Another point defect in case of pure metal is known as self-interstitial defect. When an extra atom of the same species occupies any of the interstitial voids in the crystal, it is known as interstitial defect. This results in a high amount of distortion, as the size of the interstitial free space is significantly smaller than the size of the atom. Hence, these interstitials are generally not produced naturally, rather due to external events like irradiation. Figure 3.3 depicts the difference between vacancy and interstitial defect in crystals.

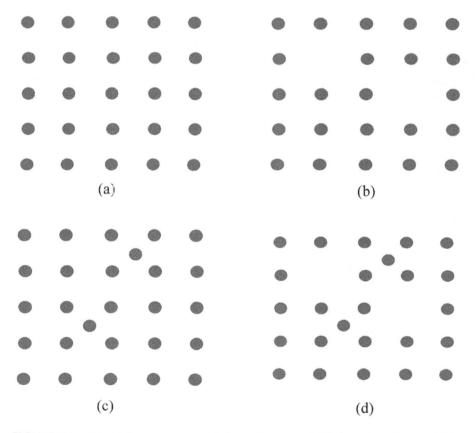

FIGURE 3.3 Point defects in pure metal: (a) perfect crystal, (b) vacancies, (c) interstitials, and (d) equal number of vacancies and interstitials.

Defects in Crystalline Solids 41

3.2.1.3 Point Defect in Ionic Compound

Analysis of point defect in the case of ionic compound is complex, as charge neutrality is an essential criterion in this regard. The ratio of the number of missing cations and anions should be such that the overall charge is balanced. For simplicity, let us assume an ionic compound where both cation and anion are having same valency such that the formula for the compound is MX (M^{n+} is the cation and X^{n-} is the anion).

When an equal number of vacancies are created on M (or X) sublattice and the same number of interstitial M (or X) are formed without interruption of X (or M) sublattice, the defect is known as Frankel defect. When an equal number of vacancies are present both in M and X sublattices, the defect is termed as Schottky defect. This is shown in Figure 3.4.

3.2.1.4 Extrinsic Point Defect

Practically, it is impossible to obtain a perfect pure metal (100% purity) without any impurity or foreign atoms. These foreign atoms constitute extrinsic defect in the host metal, and if the foreign atom gets dissolved in the parent metal, it is known as solid solution. Depending on the position of the foreign atom in the parent metal, it is further classified into two classes. If the foreign atom occupies the interstitial void

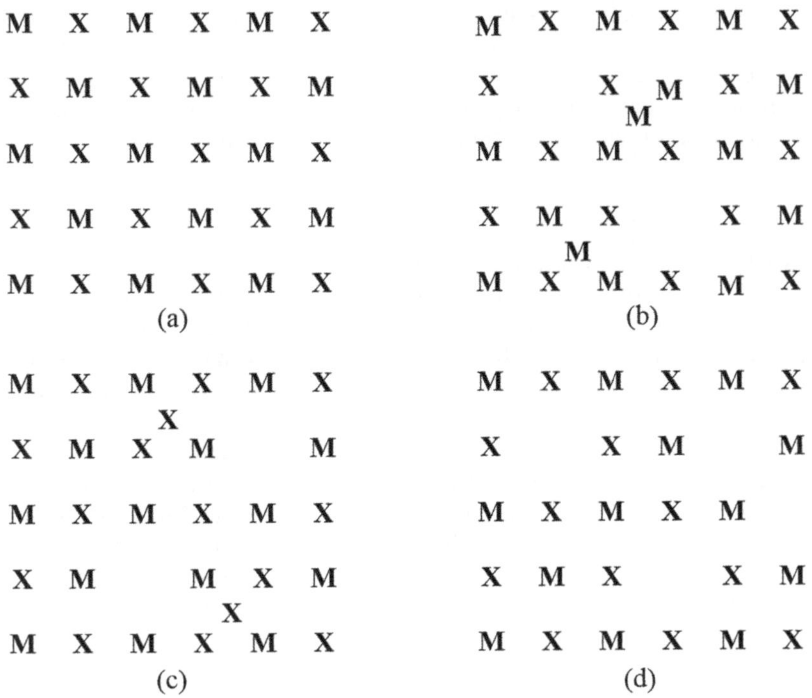

FIGURE 3.4 Point defects in ionic compound structure: (a) perfect crystal structure, (b and c) Frankel defect, and (d) Schottky defect.

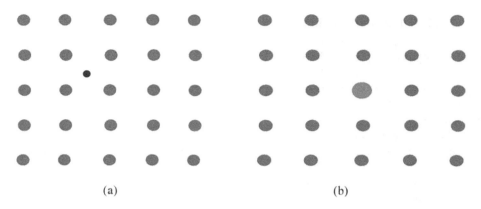

FIGURE 3.5 Solid solution: (a) interstitial and (b) substitutional.

position in the metal, then it is referred to interstitial solid solution, and if the foreign atom occupies the lattice position of the parent metal (in other words, replaces one of the parent atoms from its position), it is known as substitutional solid solution. An excellent example of interstitial solid solution is steel where the solute foreign atom (carbon in this case) occupies the interstitial position between the parent iron atoms. Typically, when the foreign atom is of very small size in comparison to the parent metal atom, interstitial solid solution is formed. In the case of substitutional solid solution, both the parent and the foreign atoms should have almost the same atomic size. For example, Cu atom can substitute Ni atom without disturbing the crystal arrangement of Ni lattice forming a substitutional solid solution. A schematic of the interstitial and substitutional solid solution is shown in Figure 3.5.

3.2.2 One-Dimensional or Line Defect

In crystalline solids, line defect refers to those defects producing lattice distortion around a line. This class of defect is very important as far as mechanical properties of the material are concerned. The one-dimensional or line defect in case of crystalline solids is known as dislocation. Dislocations may be produced during solidification or during plastic deformation or by vacancy condensation. Presence of dislocations allows plastic deformation of the material far below the fracture stress of the material without disturbing the crystal structure. **Dislocation line** is defined as the boundary between the disturbed and undisturbed portion of the crystal. Another important characteristic of dislocation is the **Burgers vector**. The Burgers vector is the vector having slip direction, and magnitude is the interatomic distance between two consecutive atoms along the slip direction. Based on the geometry of dislocation, it is divided into two groups: (i) edge dislocation and (ii) screw dislocation.

3.2.2.1 Edge Dislocation

When an extra portion of an atomic plane is present (or in other words, a portion of an atomic plane is missing) in a crystal, the defect formed is known as edge

Defects in Crystalline Solids

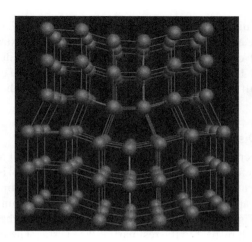

FIGURE 3.6 Schematic of edge dislocation.[1]

dislocation. This extra plane is known as extra half plane. A schematic diagram of edge dislocation is shown in Figure 3.6.

To get an idea about the formation of edge dislocation in crystal, consider Figure 3.7. Application of small shear force to a perfect crystal causes localized plastic deformation on the slip plane (Figure 3.7a).

Because of the local deformation, the crystallographic arrangement is disturbed in that vicinity, and formation of an extra half plane takes place (Figure 3.7b). Dislocation line in case of edge dislocation is the edge of the extra half plane shown as line AB. With increase in the magnitude of the applied shear force, the dislocation line moves toward left, i.e., perpendicular to itself. The Burgers vector in this case is shown in Figure 3.7b, which has existence on the slip plane along the slip direction. Hence, in case of edge dislocation, the Burgers vector is perpendicular to the dislocation line (from Figure 3.7).

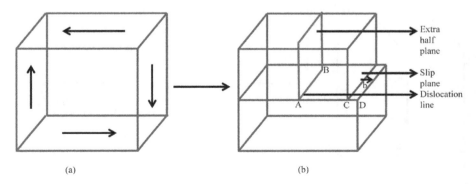

FIGURE 3.7 Formation of edge dislocation: (a) perfect crystal and (b) crystal with an edge dislocation.

A positive edge dislocation is represented by an inverted "T" symbol (⊥) where the extra half plane exists at the top of the crystal. When the extra half plane exists at the bottom part of the crystal, it is referred to negative edge dislocation and represented as (⊤). In case of an edge dislocation, there exists a compressive stress, where the extra half plane is present, and a tensile stress in the other part of the dislocation line.

3.2.2.2 Screw Dislocation

When a part of the crystal has a relative angular orientation with respect to the other part, the defect is referred as screw dislocation. The word screw is derived from the spiral nature of the atomic arrangement around the dislocation line.

Because of the applied shear stress, a part of the crystal undergoes deformation as shown in Figure 3.8. A portion of the upper part of the crystal has been moved partially with respect to the rest part of the crystal. The line XY represents the boundary between deformed and undeformed portions of the crystal; hence, it is the dislocation line. With a higher applied shear stress, the boundary moves toward left. The Burgers vector is shown in Figure 3.8b, which represents the direction of slip. From the figure, it is very clear that in case of a screw dislocation, the Burgers vector is parallel to the dislocation line unlike in case of edge dislocation. Similar to that of edge dislocations, two types of screw dislocations do exist, namely positive or clockwise (↻) and negative or anticlockwise (↺) screw dislocation.

3.2.2.3 Dislocation Movement

A common mechanism by which dislocation (dislocation line) moves is known as dislocation glide. An essential criterion for dislocation glide is that, along with the movement of dislocation line, the dislocation line and the Burgers vector should always be present on the same slip plane. To understand this, again consider Figures 3.7 and 3.8. With increase in the magnitude of applied stress, the dislocation line moves in the direction of applied stress, but it always remains on the slip plane.

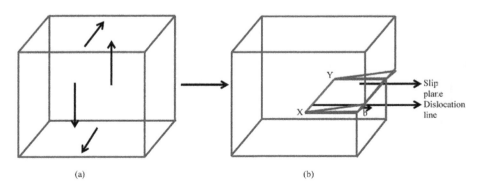

FIGURE 3.8 Formation of screw dislocation: (a) perfect crystal and (b) crystal with a screw dislocation.

Defects in Crystalline Solids

Similarly, although the magnitude of the Burgers vector increases, still it exists on the same slip plane where the dislocation line is present.

Cross-slip is a mechanism through which a screw dislocation changes its slip plane. The schematic of cross-slip is shown in Figure 3.9.

Consider a pair of slip planes ABCD and CDEF having line of intersection CD. Let us assume that there is a screw dislocation with dislocation line XY (parallel to CD) and the Burgers vector as shown in Figure 3.9 (parallel to XY) on the slip plane ABCD. With the application of force, the dislocation line moves toward CD. When the dislocation line coincides with CD, the dislocation line has now a provision to further move on the plane CDEF, as the Burgers vector at this stage will also be present on the plane CDEF fulfilling the criterion of dislocation glide on CDEF plane. This phenomenon of changing the glide plane of the dislocation is known as cross-slip. But cross-slip is only possible for screw dislocation. In case of an edge dislocation, cross-slip is not possible, which can be analyzed from Figure 3.10.

When the dislocation line will coincide with CD, the Burgers vector (perpendicular to XY and along BC) will not reside on the plane CDEF; rather, it would be at

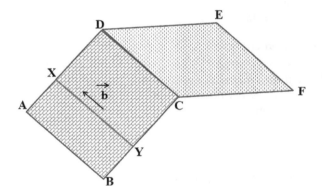

FIGURE 3.9 Schematic of cross-slip of screw dislocation.

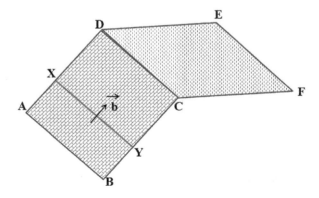

FIGURE 3.10 Schematic showing cross-slip is impossible in case of edge dislocation.

some angle to the slip plane CDEF. Hence, movement of dislocation is not possible on the slip plane CDEF. Hence, in case of an edge dislocation, cross-slip is not possible.

Another mechanism through which an edge dislocation leaves its slip plane is known as **climb**. This is related to vacancy or atomic diffusion. When a series of vacancies replace the line of atoms at the edge of the extra half plane or when a series of atomic line is added below the extra half plane (above the extra half plane in case of negative edge dislocation), the dislocation line leaves the earlier slip plane and moves to another slip plane parallel to earlier one. This phenomenon is known as dislocation climb. As climb involves diffusion, the process becomes feasible only at higher temperature.

3.2.2.4 Existence of Dislocations in Crystals

In reality, pure edge or pure screw dislocation is rare to find in crystals. Rather, most of the dislocations in crystalline materials are combination of both edge and screw dislocations. A schematic representation of a dislocation in a crystal is shown in Figure 3.11. It can be observed that the dislocation has different characters at different locations. However, the Burgers vector is invariant and hence has same magnitude and direction irrespective of the location. A dislocation also cannot be terminated inside a crystal suddenly; it can only be terminated either in itself (loop) or at a node or at the surface.

3.2.2.5 Dislocation Energy

Any dislocation has inherently a stress field associated with it. This stress is normally within the elastic limit, and thus, an elastic strain energy is developed at the dislocation. The elastic strain energy per unit length of dislocation can be determined from the following expression:

$$E = \frac{1}{2}Gb^2 \tag{3.6}$$

where G represents the shear modulus and b is the Burgers vector.

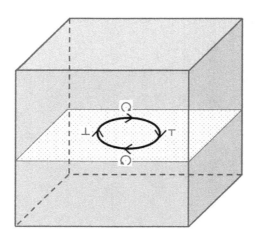

FIGURE 3.11 A schematic of existence of dislocation in a crystal.

Defects in Crystalline Solids

3.2.3 Two-Dimensional or Surface Defects

Surface defects refer to regions of crystals, where distortions lie about a surface. The thickness of the region might have dimensions around few atomic diameters. The atoms at this surface do not satisfy the minimum energy criterion, and hence, the energy of this surface is higher than that of the perfect crystal. Surface defect is classified into two classes.

3.2.3.1 External Surface Defect

The free surface of any crystalline solid represents a higher energy state. This free external surface represents the termination of the crystal. The coordination number of the atoms on this surface is less than the ideal coordination number of the crystal structure.

Consider Figure 3.12a, which represents a part of a perfect crystal, and assume that it has infinite size. Hence, every atom is surrounded by four nearest neighbor atoms in 2-D. To make a free surface, one has to cut into two parts along the line AB, as shown in Figure 3.12b. To cut, all the bonds along that line are to be broken which needs some energy, and that energy is now stored on both the surfaces, known as surface energy. Thus, the coordination numbers of corresponding atoms whose bonds are broken possess a lower coordination number less than the ideal one. Hence, the free surface is a defect.

The free surface being of higher energy acts as catalyst for many reactions. Thus, the process of corrosion starts from the free surface. Powder samples (higher free surface area) thus are more prone to chemical reactions. During sintering of green pellets, this high surface energy is accounted for the agglomeration of the fine powder particles to reduce the free surface area and thus net reduction in surface energy.

3.2.3.2 Internal Surface Defect

This refers to surface defect that exists inside the solid. The most frequently encountered internal surface defect is grain boundary. The interface between neighboring grains is termed as grain boundary. During solidification of a polycrystalline material from its molten liquid bath, several nuclei are formed and they grow simultaneously in their own crystallographic directions, and later they start touching each other toward the end of solidification. Now, the atoms exactly at this interface possess high energy, as they deviate from the ideal crystallographic arrangement.

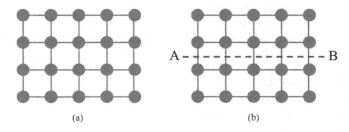

FIGURE 3.12 Free surface as a defect: (a) perfect crystal and (b) cutting along AB to make two free surfaces.

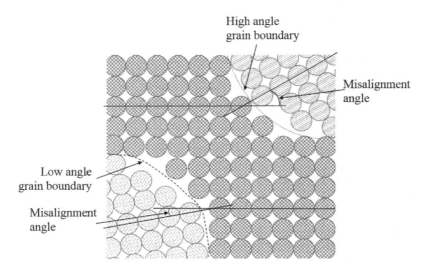

FIGURE 3.13 Grain boundaries in polycrystalline metal.

This orientation difference between two consecutive grains gives rise to grain boundary defect. Although the structure of each gain is same (the values of a, b, c and α, β, γ for each grain are the same), but the crystallographic axes of each grain make some angle with the other, e.g., the "a" axes of grain 1 and grain 2 are not unidirectional as shown in Figure 3.13. The angle between these two axes ("a" axis of grain 1 and "a" axis of grain 2), known as orientation difference represents the angle of the grain boundary.

The boundary between two adjacent grains is generally distorted and less packed as that of bulk grains. The width of this zone is quite narrow and ranges up to five atomic diameters. Due to less packing of the grain boundary, it exhibits a higher energy than the grain. The extent of this energy further increases with more distortion, i.e., more free space.

The orientation mismatch between two adjacent grains is defined as the misalignment angle. Grain boundaries are further divided into two groups depending on the misalignment angle. A low-angle grain boundary (LAGB) is defined as the grain boundary with a misalignment angle less than 15°. These grains are also known as subgrains. The LAGBs are generally formed by an array of dislocations. If the LAGB is formed by edge dislocations, it is known as **tilt boundary**, and if it is formed by screw dislocations, it is known as **twist boundary**. In contrast, in case of a high-angle grain boundary, the grains are misoriented by an angle more than 15°.

3.2.3.3 Twin Boundary

Another example of two-dimensional defect is twins or twin boundary. A twin boundary is defined as the boundary between a part of the grain and its mirror image as shown in Figure 3.14. Generally, a twin boundary exhibits much lower energy than that of grain boundary. Twin is again divided into two categories: (i) mechanical/deformation twin and (ii) annealing twin. Deformation twins are

Defects in Crystalline Solids

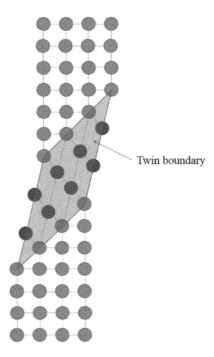

FIGURE 3.14 A schematic representation of twin boundary.

generally evolved during high strain rate cold working of the metals having few slip systems. It happens when a part of the crystal is sheared by a specific magnitude such that it becomes the mirror image of the remaining part of the crystal. Out of the common crystal structures, deformation twins are more prominent in hexagonal close-packed (HCP) (zinc, magnesium, etc.) and tetragonal (tin) structures due to availability of less slip systems.

Annealing twins are mostly formed in face-centered cubic (FCC) metals during recrystallization of deformed metals due to growth accidents. It is quite prominent in copper, nickel, α-brass, and austenitic iron.

3.2.3.4 Stacking Fault

As the name suggests, it is the fault in the stacking sequence of a crystalline solid. For example, the stacking sequence of an FCC crystal is …ABCABCABC…. However, at instances, the stacking sequence of a part of the crystal may be in the form of …ABCACABC…, due to some missing plane. Now, in this part of the crystal, the stacking CACA represents an HCP system, but in reality, it is the stacking fault in the FCC crystal. In another example, the stacking sequence of the same FCC crystal may change to ABCACBCABC, due to existence of an additional plane. In this case, ACB forms a twin. Therefore, this kind of stacking fault is also referred as microscopic twin in FCC metals. The stacking fault energies (SFEs) of metals normally lie in the range of 0.01–0.1 J/m^2. A lower SFE metal has wider stacking fault, and thus, strain hardens rapidly. SFEs of some common metals and alloys are provided in Table 3.1.

TABLE 3.1

SFEs of Some Common Metals and Alloys

Metal/Alloy	Stacking Fault Energy (mJ/m²)
Brass	<10
303 stainless steel	8
304 stainless steel	20
310 stainless steel	45
Silver	~25
Gold	~50
Copper	~80
Magnesium	125
Nickel	~150
Aluminum	~200

SFEs, stacking fault energies

3.2.4 THREE-DIMENSIONAL OR VOLUME DEFECT

Any defect having a three-dimensional structure is a volume defect. In metals, various commonly observed volume defects are pores/voids, crack, impurities, and so on. Pores/voids are normally introduced into the material during the fabrication stage. Cracks are formed due to localized stress, which may be formed during handling or during service condition. Both pores and cracks act as stress concentrators and thus decrease the strength of the materials. Impurities also may be entrapped in the metal during its processing, but in some cases, impurities are intentionally introduced into the metals. These impurities sometimes obstruct the motion of the dislocation and thus increase the strength of the metal. Such types of impurities are termed as dispersoids. Sometimes, a new phase like precipitate may also be formed in the material as a result of phase transformation, which also improves the strength of the material by the precipitation hardening mechanism.

REFERENCE

1. Crystal Defects: Linear Defects (Dislocations). https://www.nde-ed.org/ EducationResources/CommunityCollege/Materials/Structure/linear_defects.htm.

FURTHER READING

Dieter, G. E. *Mechanical Metallurgy*. (McGraw-Hill, New York, 1988).
Abbaschian, R. & Reed-Hill, R. E. *Physical Metallurgy Principles*. (Cengage Learning, Massachusetts, 2008).

4 Solid Solutions

4.1 INTRODUCTION

A solution is basically a homogeneous mixture of one or more types of solutes in a solvent. A solvent is further defined as the component where a solute is dissolved, and a solute is the entity that is dissolved in the solvent to obtain the homogeneous mixture. Solute and solvent may be either solid or liquid or gas. Homogeneous in this context is generally treated as a single-phase substance having uniform physical, chemical, and mechanical properties. As per the nature of the solvent, a solution is classified as either solid or liquid or gaseous solution. The maximum limit of the weight of the solute that can be dissolved in a solvent is termed as solubility and is usually a function of temperature and pressure. Depending on this, a solution may be termed as unsaturated, saturated, or supersaturated. For most of the solutions, the solubility limit increases with increase in temperature. For example, the solubility of sugar increases in water upon heating. Furthermore, the solubility limit generally is higher in the liquid state than that of the solid state. In metallurgy, the focus is usually on solid solutions/alloys, where the solvent is primarily a solid-state material at ambient environment, e.g., steel is a solid solution of carbon in iron. Here, carbon acts as the solute, whereas iron behaves like the solvent. In case of a solid solution, the crystallographic arrangement of atoms of the solvent phase remains unaltered by the process of solution formation. In context to solutions, we usually encounter the term solubility. Solubility is defined as the maximum amount of the solute that can be dissolved in a solvent at a given temperature and pressure combination. It comes from the concept of free energy change, i.e., a solution is evolved only when there is a decrement in the net free energy of the system. The properties of solid solutions to a large extent are dependent on their composition. In addition, the type of phases and their morphologies may also influence the bulk performance of the solid solution to a significant degree.

4.2 TYPES OF SOLID SOLUTIONS

Solid solutions are classified into two groups based on the atomic arrangement of the solute atoms in the solvent lattice.

4.2.1 INTERSTITIAL SOLID SOLUTION

An interstitial solid solution is said to be formed only when the solute atoms occupy the interstitial free space among the atoms of the solvent crystal lattice. Usually, elements with very small atomic size (less than 1 Å) are suitable for having an interstitial solid solution, which can fit into the available interstitial sites in the solvent phase. Hydrogen, carbon, boron, and nitrogen are among the solutes that can be accommodated at such interstitial sites. However, hydrogen (atomic size ~0.46 Å) is the only

51

element that has a size usually lower than that of the interstitial site. Other elements, being relatively larger than the interstitial sites, cause a volume expansion of the lattice. Therefore, only a limited number of metals (such as Fe, Ni, and Ti) do have significant solubility for these elements. Steel is a classic example of interstitial solid solution/alloy, where the carbon atoms do occupy the interstitial sites of iron lattice.

The simplest form of difference between an interstitial alloy and interstitial compound is the concentration of solute. The usual concentration of solute in an interstitial alloy is much lesser than that of the interstitial compound. With a large difference in the atomic radii of solute and solvent, the mobility of the solute atoms is significant at lower solute concentration. With increase in the solute concentration, the solution approaches toward saturation, and eventually, the mobility of the solute atoms is lowered. Beyond saturation, the mobility of the solute atoms is drastically reduced, and formation of interstitial compound having a composition stoichiometry starts forming. The concentration of both the solute and the solvent is fixed in case of an interstitial compound, and thus, it has a fixed chemical formula. On the contrary, an interstitial solid solution has a range of composition.

Coming to the solubility of interstitial alloy, as the solute atoms occupy only the free space available in the solvent lattice, the solubility of interstitial solute in such alloy is limited. With phase transformation and temperature, this solubility limit may further be altered depending on the volume of free space and size of the interstitial voids. For example, the maximum solubility of carbon in α-iron is 0.025% (at 723°C), whereas in the case of γ-iron, the solubility limit is 2.06% (at 1147°C).

4.2.2 Substitutional Solid Solution

If the solute atoms substitute/replace the atoms of the solvent phase without destroying the crystal structure of the solvent in a solution, it is termed as substitutional solid solution. Daily life examples of substitutional alloys include brass which is comprised of Cu and Zn, and gold–silver alloys used for making jewelry. Au (or Ag) may replace FCC Ag (or Au), resulting a FCC alloy. A schematic representation of pure metal, interstitial alloy and substitutional alloy is shown in Figure 4.1. In contrast to the interstitial solid solutions, the solubility of substitutional solid solutions is mostly higher. However, there are several factors that combinedly govern the solubility of an

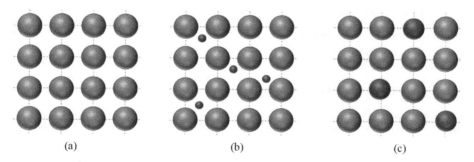

FIGURE 4.1 Arrangement of atoms in a (a) pure metal, (b) interstitial alloy, and (c) substitutional alloy.

Solid Solutions

alloy system. For complete solubility, several conditions must be met as formulated by English metallurgist and materials scientist William Hume-Rothery and known as the Hume-Rothery rules.

- Crystal structure: If both the constituents exhibit the same crystal structure, then the chance of having complete solubility is higher.
- Relative size difference: For achieving a substantial solubility, the relative atomic size difference between the solvent and solute must not exceed 15%.

$$\text{Mismatch} = \frac{r_{solute} - r_{solvent}}{r_{solvent}} \times 100\% < 15\%$$

- Chemical affinity: There should not exist a strong chemical affinity between the solute and the solvent in order to have a higher solubility. It can also be expressed in terms of electron negativity. There is a higher chance of compound formation with solute–solvent combination having a higher difference in electron negativity. For maximum solubility, ideally the difference between the electron negativity of solute and solvent should be close to 0.
- Valency: A metal can be dissolved in a metal of lower valence to a larger extent than that of higher valence. For example, Al can be dissolved in Ni to a higher extent than Ni in Al. For having a high solubility, typically both the solute and the solvent should have similar valency.

For complete solubility, all these aforementioned criteria must be met. An excellent example of complete solubility is copper–nickel alloy system. Both Cu and Ni crystallize in FCC structure, and the difference in their atomic radii is less than even 1%. Both Cu and Ni have very close electron negativity and valency. All these factors together make it possible to have a Cu–Ni alloy system having complete solubility in each other. However, it is fairly easy to check whether a particular combination of metals exhibits favorable size difference. This only factor is sufficient to restrict the solubility to a very low limit. Upon fulfillment of the favorable size conditions, other criteria must also be considered to predict the solubility of the alloy system.

4.3 ELECTRON-TO-ATOM RATIO

This is somehow related to the valency factor as discussed in the Hume-Rothery rules of solubility. The electron-to-atom ratio (e/a) is an important parameter which influences the solubility of an alloy system. The solute that is having a tendency to increase the overall e/a of the alloy is expected to have a higher solubility. As per the earlier example, Cu (higher valency) has a higher solubility in Ni (lower valency) than the reverse. Similarly, Mg (+2) has 70 atom% solubility in Li(+1), whereas Li has only 24.5 atom% solubility in Mg. Interestingly, in the case of copper- and silver-based alloys, the solubility limit of many elements typically corresponds to a value where the e/a approaches 1.35–1.4, which corresponds to the critical concentration of 1.36 at which the Fermi sphere tends to reach the first Brillouin zone. For example, the maximum solubility of Zn in Cu is 42 atom%, i.e., in a representative alloy of 100 atoms, there will be 42 atoms of Zn and thus 58 atoms of Cu. Corresponding to the valency of Zn (+2) and Cu (+1), the total number of valency electron thus is 142 (= 42×2+58×1). With this, the e/a ratio becomes ~1.4.

4.4 ENTHALPY OF FORMATION OF A SOLID SOLUTION

Enthalpy is the measure of the heat/internal energy change during any reaction/solution/transformation. In case of a solid solution, there may be release of heat or absorption of heat or no change in heat due to the mixing process. For an ideal solution, the net change in enthalpy should be 0, i.e., $\Delta H_{mix} = 0$. However for nonideal solutions, (i) $\Delta H_{mix} < 0$, which indicates an exothermic reaction or heat is evolved during the reaction, or (ii) $\Delta H_{mix} > 0$, which indicates an endothermic reaction or heat is absorbed during the reaction. Enthalpy of mixing depends on the likeliness of an atom toward similar kind of atoms or dissimilar atoms. In an alloy system of A and B, there may exist three different kinds of bonds, A–A, B–B, and A–B as shown in Figure 4.2.

The number of A–A, B–B, and A–B bonds depends on the concentration and distribution of the solute atoms in the solution. If there are M_{AA} number of A–A bonds, M_{BB} number of B–B bonds, and M_{AB} number of A–B bonds, then the internal energy of the A–B solid solution (E_{ss}) may be represented as,

$$E_{ss} = M_{AA}E_{AA} + M_{BB}E_{BB} + M_{AB}E_{AB} \tag{4.1}$$

where E represents the bond energy, and the subscripts AA, BB, and AB denote the respective bonds. In the A–B alloy, let us now consider that there are N_A number of A atoms and N_B number of B atoms, and each atom is having coordination number of z. The total number of bonds involving A atoms is

$$N_A z = 2M_{AA} + M_{AB} \tag{4.2}$$

Thus,

$$M_{AA} = \frac{N_A z - M_{AB}}{2} \tag{4.3}$$

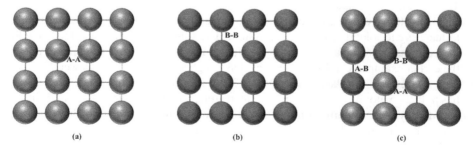

FIGURE 4.2 Atomic bonding in (a) A lattice, (b) B lattice, and (c) A–B lattice (solid solution).

Solid Solutions   55

Similarly,

$$N_B z = 2M_{BB} + M_{AB} \tag{4.4}$$

And,

$$M_{BB} = \frac{N_A z - M_{AB}}{2} \tag{4.5}$$

Now the expression for internal energy becomes

$$E_{ss} = \frac{N_A z - M_{AB}}{2} E_{AA} + \frac{N_A z - M_{AB}}{2} E_{BB} + M_{AB} E_{AB}$$

$$E_{ss} = \frac{N_A z}{2} E_{AA} + \frac{N_B z}{2} E_{BB} + \left(E_{AB} - \frac{E_{AA} + E_{BB}}{2} \right) M_{AB} \tag{4.6}$$

The net energy of pure components A and B before formation of solid solution can be presented as

$$E_1 = \frac{N_A z}{2} E_{AA} + \frac{N_B z}{2} E_{BB} \tag{4.7}$$

Hence, the enthalpy change due to formation of solid solution is

$$\Delta H_{mix} = E_{ss} - E_1 \tag{4.8}$$

$$\Delta H_{mix} = \left(E_{AB} - \frac{E_{AA} + E_{BB}}{2} \right) M_{AB} \tag{4.9}$$

The previous expression indicates that

a. For an ideal solution, $\Delta H_{mix} = 0$, or $E_{AB} = \dfrac{E_{AA} + E_{BB}}{2}$. The bond energy A–B is the average of the bond energies A–A and B–B.
b. For an exothermic reaction, $\Delta H_{mix} < 0$, or $E_{AB} < \dfrac{E_{AA} + E_{BB}}{2}$. A atoms want to be surrounded by B atoms and vice versa.
c. For an endothermic reaction, $\Delta H_{mix} > 0$, or $E_{AB} > \dfrac{E_{AA} + E_{BB}}{2}$. Both A and B atoms want to be surrounded by similar atoms.

However, when E_{AB} and $\dfrac{E_{AA} + E_{BB}}{2}$ values are close to each other, a random arrangement of atoms may be expected in the A–B alloy leading to a regular solution. In such cases, a good approximation of enthalpy of mixing is

$$\Delta H_{mix} = X_A X_B \Omega \tag{4.10}$$

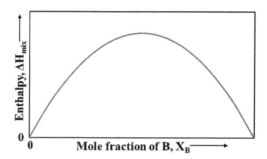

FIGURE 4.3 Enthalpy of mixing vs. mole fraction of B in an A–B solid solution.

$X_A = \dfrac{N_A}{N_T}$, or mole fraction of A.

$X_B = \dfrac{N_B}{N_T}$, or mole fraction of B.

N_T is the total number of atoms.

$$\Omega = \left(E_{AB} - \frac{E_{AA} + E_{BB}}{2} \right) \frac{zN_T}{2} \qquad (4.11)$$

The expression for ΔH_{mix} may further be rewritten as

$$\Delta H_{mix} = (1 - X_B) X_B \Omega = \left(X_B - X_B^2 \right) \Omega \qquad (4.12)$$

For a positive Ω, the variation in the enthalpy of mixing is plotted in Figure 4.3 as a function of mole fraction of B.

4.5 ENTROPY OF FORMATION OF A SOLID SOLUTION

Entropy is conventionally defined as the degree of randomness of a system. A mathematical expression for the change in entropy (S) can be presented by the Boltzmann equation mentioned in the following:

$$\Delta S = k_B \ln \omega \qquad (4.13)$$

Where k_B is Boltzmann's constant and ω is typically known as number of possible distinguishable microstates or arrangements.

Generally, entropy has two contributions for a solid solution. The first one is the thermal entropy that comes from the thermal energy of the system. Secondly, the different ways that the atoms of solvent and solute arrange themselves to form the solution are known as configurational entropy. Consider one mole of a solid solution, i.e., N_0 having N_A number of "A" atoms and N_B number of "B" atoms.

$$N_0 = N_A + N_B \qquad (4.14)$$

Solid Solutions

In the crystal structure of the solid solution, there are now total N_0 lattice sites, which are to be occupied by N_A number of A atoms and N_B number of B atoms. Let us now fill the lattice by A and B atoms. The first A atom now can occupy any of the N_0 sites, the second can occupy the rest $N_0 - 1$ sites, and the third then can occupy the remaining $N_0 - 2$ sites. In this way, the last A atom can occupy any of the available $N_0 - N_A + 1$ sites. As all the A atoms are similar, the total number of ways that the N_A number of A atoms can fill the lattice can be represented as

$$\omega_A = \frac{N_0 \times (N_0 - 1) \times (N_0 - 1) \times (N_0 - 1) \times \cdots \times (N_0 - N_A + 1)}{N_A!} \qquad (4.15)$$

In a similar way, the first atom of B can occupy any of the $N_0 - N_A$ available free sites, the second atom of B can occupy any of the $N_0 - N_A - 1$ available free site and so on, and the last B atom will occupy the last available free lattice site. In this way, the total number of possible ways that the N_B number of B atoms can fill the lattice can be represented as

$$\omega_B = \frac{(N_0 - N_A) \times (N_0 - N_A - 1) \times (N_0 - N_A - 2) \times \cdots \times 3 \times 2 \times 1}{N_B!} \qquad (4.16)$$

Now combining both A and B atoms, the total number of ways that these N_A number of A atoms and N_B number of B atoms can fill the crystal with N_0 lattice points is

$$\omega = \omega_A \times \omega_B$$

$$\omega = \frac{N_0 \times (N_0 - 1) \times (N_0 - 1) \times (N_0 - 1) \times \cdots \times (N_0 - N_A + 1)}{N_A!}$$

$$\times \frac{(N_0 - N_A) \times (N_0 - N_A - 1) \times (N_0 - N_A - 2) \times \cdots \times 3 \times 2 \times 1}{N_B!}$$

$$\omega = \frac{N_0!}{N_A! \times N_B!} \qquad (4.17)$$

Following equations 4.13 and 4.17,

$$\Delta S_{\text{mix}} = k_B \ln \left(\frac{N_0!}{N_A! \times N_B!} \right)$$

$$\Delta S_{\text{mix}} = k_B \left(\ln N_0! - \ln N_A! - \ln N_B! \right)$$

Using Sterling's approximation, i.e., $\ln(x!) \approx x \ln x - x$

$$\Delta S_{\text{mix}} = k_B \left(N_0 \ln N_0 - N_0 - N_A \ln N_A + N_A - N_B \ln N_B + N_B \right)$$

$$\text{As } N_0 = N_A + N_B$$

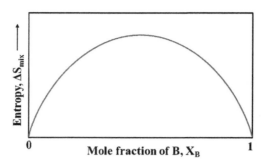

FIGURE 4.4 Variation in entropy of mixing as a function of composition of the solid solution.

$$\Delta S_{mix} = k_B \left(N_0 \ln N_0 - N_A \ln N_A - N_B \ln N_B \right)$$

Again as $N_A = N_0 X_A$ and $N_B = N_0 X_B$,

$$\Delta S_{mix} = k_B \left(N_0 \ln N_0 - N_0 X_A \ln(N_0 X_A) - N_0 X_B \ln(N_0 X_B) \right)$$

$$\Delta S_{mix} = k_B N_0 \left(\ln N_0 - X_A \ln(N_0 X_A) - X_B \ln(N_0 X_B) \right)$$

$$\Delta S_{mix} = k_B N_0 \left(\ln N_0 - X_A \ln N_0 - X_A \ln X_A - X_B \ln N_0 - X_B \ln X_B \right)$$

$$\Delta S_{mix} = k_B N_0 \left(\ln N_0 - X_A \ln N_0 - X_B \ln N_0 - X_A \ln X_A - X_B \ln X_B \right)$$

$$\Delta S_{mix} = k_B N_0 \left(\ln N_0 (1 - X_A - X_B) - X_A \ln X_A - X_B \ln X_B \right)$$

As $k_B N_0 = R$ (universal gas constant) and $X_A + X_B = 1$,

$$\Delta S_{mix} = -R(X_A \ln X_A + X_B \ln X_B) \quad (4.18)$$

Equation 4.18 indicates that ΔS_{mix} is positive, as the mole fractions X_A and X_B are less than unity, and logarithm of these values must be negative. A schematic representation of the entropy of mixing of a solid solution as a function of composition is shown in Figure 4.4.

4.6 FREE ENERGY CHANGE UPON FORMATION OF A SOLID SOLUTION

As per Gibb's free energy, the free energy of formation of a solution can be represented by the following expression:

$$\Delta G_{mix} = \Delta H_{mix} - T\Delta S_{mix} \quad (4.19)$$

In case of an ideal solution (i.e., $\Delta H_{mix} = 0$), the free energy of formation of a solid solution is solely governed by the entropic contribution, i.e.,

Solid Solutions

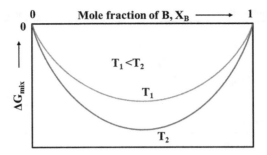

FIGURE 4.5 Effects of composition of the solution and temperature on the free energy change for an ideal solution.

$$\Delta G_{mix}^{ideal} = -T\Delta S_{mix} \quad (4.20)$$

$$\Delta G_{mix}^{ideal} = RT(X_A \ln X_A + X_B \ln X_B) \quad (4.21)$$

Hence, the free energy of an ideal solution seems to be more negative, or the process of solution formation is more spontaneous when the temperature increases, which can also be seen from Figure 4.5.

If we consider the actual free energy (not free energy change) of the solution before (G_1) and after (G_2) mixing (Figure 4.6),

$$G_1 = G_A X_A + G_B X_B \quad (4.22)$$

$$G_2 = G_1 + \Delta G_{mix}^{ideal} = G_A X_A + G_B X_B + RT(X_A \ln X_A + X_B \ln X_B) \quad (4.23)$$

For real solutions, although the second component (i.e., $T\Delta S_{mix}$ or entropic contribution) of equation 4.19 is negative, ΔH_{mix} may be either positive or negative. Thus, ΔG_{mix}^{real} may be positive or negative. The expression for ΔG_{mix}^{real} may be represented as

$$\Delta G_{mix}^{real} = X_A X_B \Omega + RT(X_A \ln X_A + X_B \ln X_B) \quad (4.24)$$

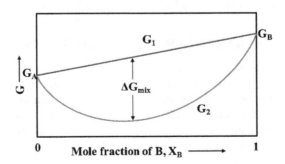

FIGURE 4.6 Free energy of the solution before and after mixing, and free energy of mixing as functions of temperature.

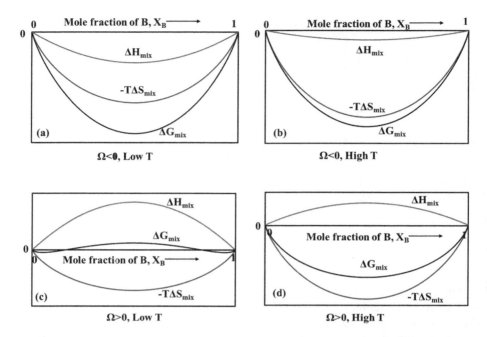

FIGURE 4.7 Enthalpy, entropy, and free energy change upon solid solution formation at various Ω and T combinations.

If formation of solid solution is as exothermic process, i.e., $\Delta H_{mix} < 0$, ΔG_{mix}^{real} will be always negative irrespective of temperature and composition; however, the magnitude of the free energy change will depend on both temperature and composition of the solid solution as presented in Figure 4.7a and b. In case it is an endothermic reaction, it will be difficult to confirm whether it will yield a positive or negative free energy change. For a fixed composition, ΔG_{mix}^{real} will be negative at relatively higher temperature, as the magnitude of the entropic contribution will be higher than that of enthalpic contribution (Figure 4.7d). At relatively lower temperature, the variation of ΔG_{mix}^{real} is further complex as shown in Figure 4.7c. From all these cases, it indicates that ΔG_{mix} will become negative upon addition of small amount of solute to the solvent.

4.7 ORDERED AND RANDOM SOLID SOLUTIONS

Atoms of solute and solvent in a substitutional solid solution may have random or preferred locations. In a random solid solution, all the lattice sites are identical to each other. Thus, the probability of any of the lattice site being occupied by A atom is equal to the mole fraction of A, i.e., X_A. The number and bond energies of A–A, B–B, and A–B bonds are useful in order to find out whether there is any sort of ordering in the solution. If the mean of the bond energies of A–A and B–B bonds $\left(\text{i.e.,} \dfrac{E_{AA} + E_{BB}}{2}\right)$ is less than E_{AB}, or in other words, if $\Omega > 0$, it indicates that A likes A more than B. Hence, A atom will be having a more number of A atoms as the

Solid Solutions

nearest neighbors than that of a random solid solution. This is known as clustering. However, when $\dfrac{E_{AA} + E_{BB}}{2}$ is greater than E_{AB} (i.e., $\Omega < 0$), A atoms have a preference toward B atoms, and thus, there may be a short-range ordering or A atoms get surrounded by more B atoms than that of the random solid solution. To know the extent of ordering in a solid solution, an ordering parameter(s) is used, which is evaluated from the following expression:

$$s = \frac{M_{AB} - M_{AB}^{ran}}{M_{AB}^{max} - M_{AB}^{ran}} \tag{4.25}$$

M_{AB} represents the number of A–B bonds in the solid solution, and the superscript ran represents the number of A–B bonds in a random solution and max represents the highest possible number of A–B bonds. A perfect random solution exhibits $s = 0$, while negative and positive values of s indicate clustering and short-range ordering in the solid solution.

4.8 INTERMEDIATE PHASES

Sometimes, at certain composition of an alloy, the crystal structure of the alloy is different from both the crystal structures of the parent elements due to the least free energy. In such cases, the evolved phase is known as an intermediate phase. The concept of lowest free energy arises from the composition of the alloy, which tends to exhibit a lower free energy than the terminal solid solutions. Various possible bonds and packing arrangement of atoms in the solid solution are very often responsible for formation of these intermediate phases.

In general, for an alloy system having a possibility of intermediate phase, generation forms a typical "U"-shaped free energy vs. composition curve near the composition of the intermediate phase. The depth and/or width of this "U"-shaped curve give information related to the composition range and stability of the intermediate phase.

In case of an intermediate phase, having a very narrow composition range with a sharp reduction in the free energy is termed as intermediate compound having almost fixed composition, i.e., A_xB_y with a stoichiometric combination of A and B as shown in Figure 4.8a. Figure 4.8b represents the existence of an intermediate phase with a broad composition variation (i.e., no stoichiometry).

Based on the type of interaction between A and B, the intermediate phase may be covalent or ionic as well. In both these cases, the intermediate phase is having a fixed composition (intermediate compound). AlSb and ZnSb are among the covalent compounds having zinc-blend structure, whereas Mn_2Sn, CoSb, and Fe_3Sn_2 are few among the covalent compounds having nickel arsenide crystal structure. Sometimes, the intermediate compound looks like a vertical straight line due to its extreme zero tolerance for additional solute and solvent concentration. There are also various ionic intermediate phases available such as MgSe, Mg_2Sn, and so on. There also exists an interstitial compound, where small atoms like hydrogen, carbon, or nitrogen occupy the interstitial voids, and formation of covalent bonds takes place. The Laves phases are also another type of intermediate phase having composition AB_2. Generally, the

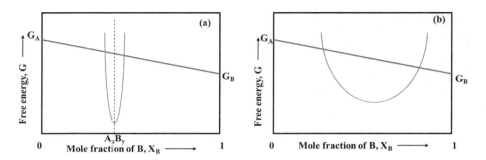

FIGURE 4.8 Free energy vs. composition diagram for (a) intermediate compound and (b) intermediate phase.

relative size difference between the atoms of solute and solvent in the case of the Laves phase is about 1.1–1.6. Atomic packing becomes most efficient if the atomic arrangement follows any of the Laves phase structure of $MgCu_2$, $MgNi_2$, or $MgZn_2$.

FURTHER READING

Hosford, W. F. *Physical Metallurgy*, Second Edition. (Taylor & Francis, Boca Raton, FL, 2010).

Porter, D. A. & Easterling, K. E. *Phase Transformations in Metals and Alloys*, Third Edition (Revised Reprint). (CRC Press, Boca Raton, FL, 1992).

5 Phase Diagrams and Phase Transformations

As discussed earlier, the existence of a particular phase of an alloy depends on the thermodynamics/free energy of the mixing process, which is altered by the temperature of the surrounding, atmospheric pressure, and composition of the alloy. To predict the existence of a possible phase or mixture of phases at certain combination of aforementioned factors, phase diagrams are useful. Phase diagrams are also referred as constitutional diagrams or equilibrium diagrams.

5.1 THERMODYNAMIC CONSIDERATIONS OF PHASE DIAGRAMS

As stated earlier, a phase diagram is basically a graph that indicates the type of phase or mixture of phases under a given circumstance. Let us consider some basic features of a phase diagram.

5.1.1 COORDINATES OF A BINARY PHASE DIAGRAM

The simplest phase diagram is a binary phase diagram having only two constituents. Hence, one possible variable here is the composition. As the metallurgical phase diagrams are constructed at normal atmospheric pressure, the environmental variable is only temperature. Hence, a binary phase diagram is constructed taking the composition of the alloy and its temperature into consideration. Conventionally, the composition (normally mentioned in terms of atom fraction or weight fraction) and temperature are presented in the abscissa (x coordinate) and ordinate (y coordinate), respectively. The weight fraction (or wt%) or atom fraction (or at%) are interchangeable based on the following expressions for an alloy of A and B.

$$\text{at.}\% \text{ of } A = \frac{\text{wt.}\% \text{ of } A}{\text{wt.}\% \text{ of } A + \text{wt.}\% \text{ of } B \times \left(\dfrac{a_A}{a_B}\right)} \times 100$$

$$\text{at.}\% \text{ of } B = 100 - \text{at.}\% \text{ of } A$$

(5.1)

where a_A and a_B are the atomic weights of A and B, respectively.

5.1.2 CONSTRUCTION OF PHASE DIAGRAM

A phase diagram is usually constructed concentrating on the solid state and early liquid state (just after melting). The traditional method of constructing a phase diagram is to cool a given composition of an alloy very slowly (to approach the equilibrium)

63

Phase Transformations and Heat Treatments of Steels

from its melt state to the desired range of temperature, which is usually the room temperature. During this whole range of cooling, the microstructure (e.g., liquid and solid with phase α and β, mixture of α and β, and so on) of the alloy is continuously monitored as a function of temperature. This process is repeated for various alloy compositions, after which the complete phase diagram is developed. In subsequent section, a detailed discussion on construction of some simple phase diagrams will be done.

5.2 GIBB'S PHASE RULE

Gibb's phase rule gives the number of possible ways to vary the controllable parameters independently in order to obtain a particular phase or phase mixture at equilibrium. Such number of ways is termed as degree of freedom (F). For being in equilibrium, the free energy of the system is minimum, and thus, there is no driving force for any transformation. In order to calculate the degrees of freedom, we first should consider the number of possible combinations among the controllable variables (compositional and environmental variables) and equalities. Consider a system containing "m" number of phases formed by "n" number of components. Let the phases be $P_1, P_2, P_3, ..., P_m$, and similarly the components are $C_1, C_2, C_3, ..., C_n$. Now let us first determine the number of compositional variables. Any phase, e.g., P_1, is comprised of $C_1, C_2, C_3, ..., C_n$. The possible ways of varying the composition of P_1 is by varying the respective percentage of $C_1, C_2, C_3, ..., C_n$. This can be done by $(n-1)$ ways, as the percentage of C_n is automatically fixed (i.e., $100 - (\%C_1 + \%C_2 + \%C_3 + \cdots + \%C_{n-1})$) after fixing the percentages of $C_1, C_2, C_3, ..., C_{n-1}$. For each of the phases $P_1, P_2, P_3, ..., P_m$, thus the number of compositional variables is $(n-1)$. So, the total number of compositional variable for all phases is $m(n-1)$.

$$\text{No. of compostional variables} = m(n-1) \tag{5.2}$$

Next coming to the environmental variables, the effects of all possible environmental variables on the stability of the phase or phase mixture should be considered. Very often, pressure and temperature are considered as the environmental variables. But it is not strictly restricted only for these two variables, and hence, other environmental factors may also be considered having significant influence on the phases.

$$\text{No. of environmental variables} = E \tag{5.3}$$

Putting all the variables together,

$$\text{Total no. of variables} = \text{No. of compostional variables}$$
$$+ \text{No. of environmental variables}$$

Hence,

$$\text{Total no. of variables} = m(n-1) + E \tag{5.4}$$

Phase Diagrams and Phase Transformations

The next thing to consider here is the number of equalities. For the phases P_1, P_2, P_3, ..., P_m to be in equilibrium, the chemical potential or vapor pressure of each of the elements/components should be the same in each of the phases. Hence,

$$\mu_{C_1}^{P_1} = \mu_{C_1}^{P_2} = \mu_{C_1}^{P_3} = \ldots = \mu_{C_1}^{P_m}$$

$$\mu_{C_2}^{P_1} = \mu_{C_2}^{P_2} = \mu_{C_2}^{P_3} = \ldots = \mu_{C_2}^{P_m}$$

$$\vdots \qquad \vdots \qquad \vdots \qquad \qquad \vdots$$

$$\mu_{C_n}^{P_1} = \mu_{C_n}^{P_2} = \mu_{C_n}^{P_3} = \ldots = \mu_{C_n}^{P_m} \tag{5.5}$$

where $\mu_{C_1}^{P_1}$ is the chemical potential of the component C_1 in the phase P_1. It can be observed from the previous expressions that, for each row of expression (i.e., involving a particular component), there are $(m-1)$ equal signs, and in expression 5.5,

$$\text{Total no. of equalities} = n(m-1) \tag{5.6}$$

The degree of freedom (DOF or F) now can be determined by subtracting the total number of equalities from the total number of variables.

$$F = m(n-1) + E - n(m-1)$$

Simplifying,

$$F = n - m + E \tag{5.7}$$

Conventionally, the number of phases, i.e., m is designated as P, and similarly the number of components, i.e., n is designated as C. If pressure and temperature are the only environmental variables, then the number of environmental variables, i.e., E becomes 2. Hence, expression 5.7 may be rewritten as

$$F = C - P + 2 \tag{5.8}$$

This expression is known as Gibb's phase rule. For most of the metallurgical practices, the number of environmental variables is 1, which corresponds to temperature, as processes are carried out at normal atmospheric pressure (which is constant). Thus, for metallurgical purpose, equation 5.8 is modified as

$$F = C - P + 1 \tag{5.9}$$

5.3 LEVER RULE

For a system comprising two phases, the Lever rule is very often used to determine the weight fraction of each phase at a given combination of temperature and alloy composition. Let us take 1 g of an alloy with composition w_B, which consists of two phases α and β at a given temperature T. Thus,

$$\text{Total weight of B in the alloy} = w_B \tag{5.10}$$

66 Phase Transformations and Heat Treatments of Steels

The weight fractions of α and β are w_α and w_β, respectively. Hence,

$$w_\alpha + w_\beta = 1 \tag{5.11}$$

Let the composition of α be w_B^α. Thus,

$$\text{Weight of B in } \alpha \text{ phase} = w_\alpha \times w_B^\alpha \tag{5.12}$$

Similarly, let the composition of β be w_B^β. And,

$$\text{Weight of B in } \beta \text{ phase} = w_\beta \times w_B^\beta = \left(1 - w_\alpha\right) \times w_B^\beta \tag{5.13}$$

The total amount of B in the alloy must be equal to the sum of the B in α and β phases. Accordingly,

$$w_B = w_\alpha \times w_B^\alpha + \left(1 - w_\alpha\right) \times w_B^\beta$$

Simplifying and rearranging,

$$w_\alpha = \frac{w_B^\beta - w_\beta}{w_B^\beta - w_B^\alpha}$$

and $\tag{5.14}$

$$w_\beta = 1 - w_\alpha = \frac{w_\beta - w_B^\alpha}{w_B^\beta - w_B^\alpha}$$

Applications of the Lever rule in various phase diagrams are discussed later.

5.4 TYPES OF PHASE DIAGRAMS AND PHASE TRANSFORMATIONS

There exist several types of phase diagrams depending on the solubility of the constituents of the alloy system in liquid and solid states as mentioned in Table 5.1.

5.4.1 ISOMORPHOUS PHASE DIAGRAM: BOTH COMPONENTS ARE COMPLETELY SOLUBLE IN BOTH LIQUID AND SOLID STATES

For satisfying such criteria, the Hume-Rothery rules must be obeyed. Both the metals should have the same crystal structure and have less than 8% difference in their atomic radius. Understanding the construction of the phase diagram for such system is simplest.

5.4.1.1 Construction of Type I Phase Diagram

Cooling curves are the origin of obtaining such phase diagram. It is well known that a pure metal has sharp melting/freezing point, i.e., the entire pure metal specimen

TABLE 5.1
Characteristics of Different Types of Phase Diagrams

Type	Characteristics
Type I (isomorphous)	Both the constituents are completely soluble in each other in liquid as well as solid states.
Type II (eutectic)	Both metals are completely soluble in liquid state and having zero solubility in each other in solid state.
Type III (eutectic)	Both metals are completely soluble in liquid state and having limited solubility in each other in solid state.
Type IV (congruent melting intermediate compound)	There exists an intermediate compound of fixed chemical (or very narrow range) composition that virtually divides the phase diagram into two parts. This intermediate compound is a congruent melting one.
Type V (peritectic)	In this type of phase diagram, there exists a transformation where a mixture of solid and liquid upon cooling give rise to formation of a new solid.
Type VI (monotectic)	Two liquids are partially soluble in each other.
Type VII	Two metals are insoluble in both liquid and solid states.

freezes/solidifies or melts at a particular temperature. Hence upon slow cooling, the cooling curve (temperature vs. time) of such liquid metal specimen yields a horizontal step exactly at the freezing temperature (T_m) of the metal as shown in Figure 5.1a. On the contrary, an alloy has a range of melting temperature. As shown in Figure 5.1b, the cooling curve of an alloy of given composition does not yield a horizontal step; rather, freezing starts at T_{ms} temperature and is finished at T_{mf}. These T_{ms} and T_{mf} are composition dependent, and by capturing these T_{ms} and T_{mf} for different compositions of the alloy, the entire phase diagram can be prepared.

Let us discuss this process of construction in a more detailed way. Consider an alloy system made up of A and B metals having melting points T_m^A and T_m^B, respectively, and $T_m^A < T_m^B$ (Figure 5.2).

For pure metals A and B, as there is a sharp melting point, transition from liquid to solid (or vice versa) phase takes place at that particular temperature. The degree

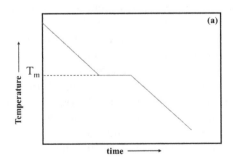
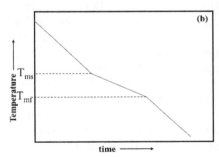

FIGURE 5.1 Cooling curves of (a) pure metal and (b) an alloy.

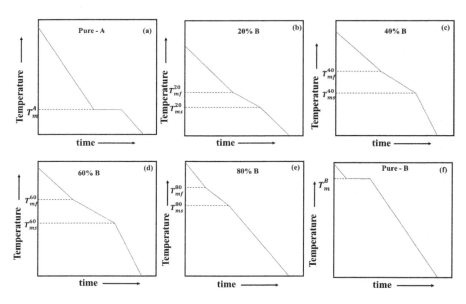

FIGURE 5.2 Schematic representation of cooling curves of (a) pure A, and A + B alloys with (b) 20% B, (c) 40% B, (d) 60% B, (e) 80% B, and (f) pure B.

of freedom at melting (where both liquid and solid phases coexist) of a pure metal is thus 0 (one component and two phases). Hence, melting of a pure metal is an invariant transformation. However, for any alloy, the DOF for melting is 1, suggesting a little change in the melting point of the metal by a small change in its composition.

5.4.1.2 Compositions of the Phases in a Phase Mixture

A type I phase diagram is generally having three distinct zones, i.e., (i) solid (single phase), (ii) liquid (single phase), and (iii) solid and liquid together (two-phase mixture). Unlike the single-phase alloy, a two-phase mixture is composed of two different phases (one solid and one liquid) of different compositions (Figure 5.3).

Assume an alloy of composition C_A is solidified from its melt (Figure 5.4). Upon cooling in the liquid state, it remains a single-phase system with the same composition. However, cooling below its liquidus temperature makes the system comprising of a mixture of both solid and liquid phases. However, the compositions of this solid and liquid in the phase mixture are different and change with temperature. The composition of these two phases at any temperature can be determined from the phase diagram by constructing a tie line (horizontal line) at that particular temperature. The intersection of the tie line with the liquidus and solidus lines gives the composition of the liquid and solid phases, respectively. In this example, when the temperature of the alloy is T_1, the solid and liquid phases are of composition C_S^1 and C_L^1, respectively (refer Figure 5.4). Further lowering in temperature to T_2 results in having a solid of C_S^2 and liquid of C_L^2 compositions as shown in Figure 5.4.

Phase Diagrams and Phase Transformations

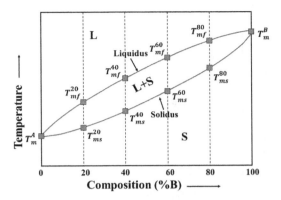

FIGURE 5.3 Construction of type I phase diagram from cooling curves.

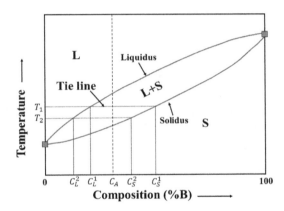

FIGURE 5.4 Concept of tie line for determination of the composition of phases.

5.4.1.3 Percentage of the Phases in a Phase Mixture

The next important thing in a phase mixture of given alloy composition is to determine the percentage of different phases that coexist with each other (Figure 5.5).

Reconsider the same alloy system, and take 100 gm of an alloy of composition C_A. At a temperature T (which corresponds to a two-phase region), let us try to determine the percentage of solid and liquid in the phase mixture. As per the concept of tie line, the compositions of the solid and liquid phases in the mixture are C_S and C_L, respectively. If we assume there is x gm of liquid, then the amount of solid will be $(100-x)$ gm. As the quantity of B (composition) of the alloy remains unchanged during the entire cooling process (mass conservation), the amount of B in the liquid state (single phase) must be equal to the sum of the amounts of B in the solid and liquid phases present in the phase mixture. Mathematically it may be represented as

$$100 \times C_A = x \times C_L + (100 - x) \times C_S$$

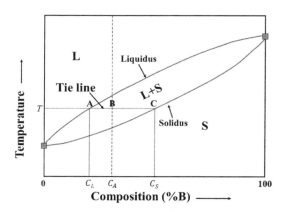

FIGURE 5.5 Determination of percentage of phases in a phase mixture by tie line.

This gives

$$x = \frac{C_A - C_S}{C_L - C_S} \text{ or } \frac{C_S - C_A}{C_S - C_L}$$

If we look at the phase diagram, $C_S - C_A$ essentially represents the length of the line BC, and $C_S - C_L$ is the length of the line AC. Hence,

$$\text{The amount of liquid (wt.\%)} = \frac{\text{Length of BC}}{\text{Length of AC}}$$

Similarly, it can also be obtained that

$$\text{The amount of solid (wt.\%)} = \frac{\text{Length of AB}}{\text{Length of AC}}$$

This is known as the Lever rule.

5.4.1.4 Properties of Type I Alloys

Very often, properties of a metal change upon alloying, and they are mostly sensitive to the composition of the alloy. The changes in these properties are mostly associated with the distortion caused to the parent crystal structure of the metal (Figure 5.6).

It can be observed from the figure that there exists a maximum in the strength vs. composition diagram, indicating the alloy to be stronger than both the pure metals. Apparently, it turns out that the alloy having 60%–70% Ni exhibits the maximum strength with an appreciable ductility.

Phase Diagrams and Phase Transformations

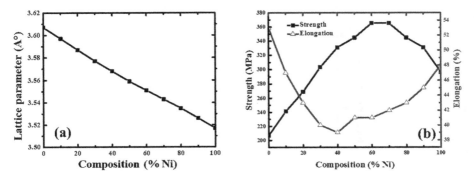

FIGURE 5.6 Changes in (a) lattice parameter and (b) strength and % elongation due to alloying.[1]

5.4.2 Eutectic Phase Diagram: Both the Constitutive Metals are 100% Soluble in Each Other in the Liquid State but Completely Insoluble or Limited Solubility in the Solid State

This is the situation when the solubility of the solute metal is negligible or limited in the solvent metal. This arises when one or more of the Hume-Rothery rules are unsatisfied. In most of the cases, the solubility is limited, and only in some rare cases, the solubility is absolute zero. In this case also, the phase diagram can be constructed by a series of cooling curves of pure terminal metals and alloys of varying compositions. As per Raoult's law, if the solute is soluble in the solvent in liquid state, but insoluble in solid state, then addition of the solute decreases the freezing point of the solvent. Hence, considering an A–B alloy system, addition of B decreases the freezing point of A, and similarly, addition of A decreases the freezing point of B. Now, it is very obvious from this statement that there exists an alloy composition corresponding to a minimum freezing point in the entire range of composition, as illustrated in Figure 5.7.

It can be noted from Figure 5.7 that upon addition of B to A, solidification starts at a lower temperature than that of pure A. However, for a broad range of the composition of the alloy, completion of the solidification is achieved at a constant temperature corresponding to T_E, known as the eutectic temperature of the alloy. Interestingly, only for a single composition (i.e., C_E) in the entire alloy system, solidification is completed only at a single temperature instead of a range. This is known as eutectic composition, where melting/solidification takes place as that of a pure metal. However, this melting cannot be termed as congruent melting as the compositions of the parent liquid and the solids formed are different. Now the phase diagram consists of four distinct zones: (i) a liquid phase (denoted by L) where both metals are completely dissolvable in each other; two phase mixtures of (ii) liquid and solid A (denoted by L + A); (iii) liquid and solid B (denoted by L + B); and (iv) solid A and solid B (denoted by A + B). Conventionally, the alloy having composition less than C_E is termed as hypoeutectic and greater than C_E is termed as hypereutectic. Let us now consider a liquid solution of eutectic composition is being solidified. The phase diagram indicates complete

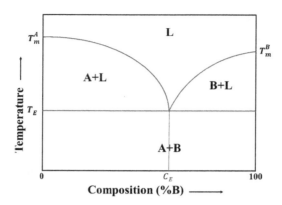

FIGURE 5.7 An eutectic phase diagram showing no solubility of the constituents in each other in the solid state.

solidification of the entire alloy at T_E temperature, which essentially means that during the entire process of solidification, the temperature of the alloy system does not change. When the temperature drops just below T_E, the liquid phase will transform into two solids represented by the end points of the horizontal strain line passing through T_E, i.e., solid A and solid B. Initially, when a tiny crystal of solid A is formed, the adjacent liquid bath becomes rich in B. To bring down the bath concentration again to C_E, a layer of solid B is formed leaving the adjacent bath with local concentration less than C_E, making conditions favorable for solidification of another layer of A. This process continues till completion of the solidification process, making the solid comprising alternate layers of pure A and pure B. This transformation is known as eutectic transformation as stated in the following:

$$\text{Liquid L} \underset{\text{Heating}}{\overset{\text{Cooling}}{\rightleftharpoons}} \text{Solid A + Solid B}$$

5.4.2.1 Microstructural Evolution During Cooling of a Eutectic Alloy

Let us consider a hypoeutectic alloy of composition C_0 being solidified from the liquid melt as shown in Figure 5.8. In the liquid stage, e.g., at temperature T_1, the alloy comprises only liquid phase having composition C_0. Upon cooling, no microstructural change takes place till the liquidus temperature. However, upon crossing the liquidus temperature, i.e., in the two-phase region, the alloy comprises both liquid (L) and solid A. In this two-phase zone, as temperature is lowered, the weight fraction of A continues to increase, and consequently, the liquid weight fraction continues to drop as per the Lever rule. As liquid amount decreases, the concentration of B in the liquid increases. For example, at temperature T_2, the system contains A in solid form and liquid L of composition C_1 (refer Figure 5.8). Just before the eutectic temperature (i.e. T_3 which is just above T_E), the system contains again pure A in solid form and liquid of eutectic composition (C_E), which can also be visualized from the phase diagram. At T_E, eutectic transformation takes place, and the liquid transforms to the eutectic mixture of solid A and solid B. Hence, at a temperature just below

Phase Diagrams and Phase Transformations

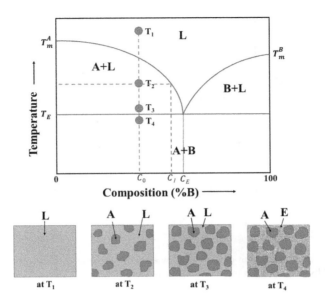

FIGURE 5.8 Microstructural evolution during solidification of a hypoeutectic alloy.

T_E, (T_4 as shown in Figure 5.8), the microstructure of the alloy contains a phase mixture of A (known as proeutectic phase) and eutectic mixture of A and B as shown in Figure 5.8. Solidification of a hypereutectic alloy can also be understood from the same principle. In such case, cooling below the liquidus temperature, crystals of solid B are formed in the liquid bath, and after the eutectic transformation, the microstructure of the alloy contains a phase mixture of proeutectic B and eutectic mixture.

There exist several eutectic alloy systems having almost negligible solubility of the constituents in each other in their solid state, such as Bi-Cd, Al-Si, and so on. Phase diagram of Bi–Cd alloy system has been presented in Figure 5.9. When the alloy of eutectic composition (i.e., 40% Cd) was cooled from liquid state, the microstructure contains 100% eutectic mixture at solid state (i.e., below 144°C) as shown in Figure 5.9 (point C). It is very obvious that this eutectic microstructure has 40% Cd and 60% Bi in it. In case of a hypoeutectic alloy, upon cooling below the liquidus temperature, proeutectic Bi is started solidifying in a liquid pool. The microstructure now consists of liquid and solid Bi (point A in Figure 5.9). Upon further cooling in this two-phase region, the amount of solid Bi increases, and gradually, the amount of liquid decreases (as from the Lever rule) and the liquid is enriched in Cd. Eventually, just above the eutectic temperature (i.e., 144°C), the liquid achieves the eutectic composition. The liquid upon crossing 144°C transforms to the solid eutectic mixture of Bi and Cd without causing any change to the proeutectic Bi. Hence, below 144°C temperature, the microstructure comprises proeutectic Bi and eutectic mixture of Bi and Cd as shown in Figure 5.9 (point B). Let us consider an alloy with 10% Cd. The Lever rule can be used to determine the weight fraction of proeutectic Bi in the solid state.

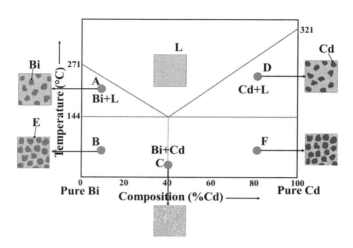

FIGURE 5.9 Phase diagram of Bi–Cd alloy system with microstructures of hypoeutectic, eutectic, and hypereutectic compositions during solidification.

$$\text{Weight fraction of proeutectic Bi} = \frac{40-10}{40} = 0.75$$

$$\text{Weight fraction of eutectic mixture} = 1 - 0.75 = 0.25$$

Hence, the composition of the alloy is 90%Bi–10%Cd, and the microstructure contains 75% proeutectic Bi and 25% eutectic mixture.

Similarly, in case of a hypereutectic alloy, proeutectic Cd and liquid are the microstructural constituents in the two-phase region as shown in Figure 5.9 (point D). Upon cooling below the eutectic temperature, now the microstructure consists of proeutectic Cd and eutectic mixture of Bi and Cd (point F in Figure 5.9). Consider a hypereutectic alloy containing 70% Cd. From the phase diagram, it can be observed that at solid state, the microstructure contains proeutectic Cd and the eutectic mixture. In the microstructure of this alloy in solid state,

$$\text{Weight fraction of pro-eutectic Cd} = \frac{70-40}{100-40} = 0.50$$

$$\text{Weight fraction of eutectic mixture} = 1 - 0.50 = 0.50$$

Another variation of eutectic alloy exists, where the constituents are having limited solubility in each other, and in reality, this is very common. A typical phase diagram of such type is shown in Figure 5.10.

As pointed out, B is having limited solubility in A. However, a solid solution is formed when the amount of B in the system remains below a critical value, and this is shown in Figure 5.10 where α is the resulted solid solution also known as terminal solid solution. Similarly, β is also another terminal solid solution of A in B. An interesting point here to note is that at eutectic temperature, B has maximum solubility

Phase Diagrams and Phase Transformations

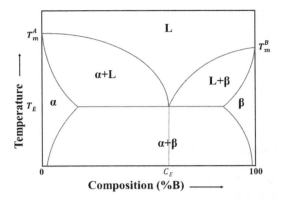

FIGURE 5.10 Eutectic phase diagram where constituents are partially soluble in each other.

in A and vice versa. Solubility here, as can be seen, is a function of temperature. As the temperature decreases below T_E, the solubility limit decreases, and the boundary between the single-phase and two-phase region below the T_E temperature is known as solvus line. In this phase diagram, also there exists a eutectic transformation. But the basic difference from the earlier one is that upon cooling the liquid of eutectic composition, it yields a eutectic mixture comprising both the terminal solid solutions (earlier one contains both pure metals).

$$\text{Liquid L} \underset{\text{Heating}}{\overset{\text{Cooling}}{\rightleftharpoons}} \text{Solid } \alpha + \text{Solid } \beta$$

It should also be noticed from the phase diagram that the composition of the eutectic mixture changes slightly, as there is a change in solubility, when cooled from T_E to room temperature.

5.4.2.2 Properties of Eutectic Alloys

It is a well-known fact that the properties of an alloy are controlled by the composition to a large extent. It certainly holds true in the case of eutectic alloys as well. Perhaps in all the eutectic alloy system, the alloy having eutectic composition exhibits the highest strength among all others. The parent metals are usually soft and plastic, whereas the eutectic phase is hard and brittle. Furthermore, as the concentration of the solute increased from pure solvent to eutectic composition, there is a linear increment in the amount of eutectic phase in the alloy. This indicates increment in strength of the alloy, as the eutectic composition is approached.

5.4.3 Peritectic Phase Diagram

If the combination of a solid and a liquid upon cooling yields a new solid, the transformation is known as peritectic reaction. Usually, the new solid formed upon this transformation is either a terminal solid solution or an intermediate phase.

$$\text{Liquid L} + \text{Solid } \alpha \underset{\text{Heating}}{\overset{\text{Cooling}}{\rightleftharpoons}} \text{Solid } \beta$$

5.5 SOME OTHER SOLID-PHASE TRANSFORMATIONS IN METALS AND ALLOYS

Apart from the transformations discussed earlier, there also exists several other phase transformations that take place in solid state only. Some of these are discussed here.

5.5.1 ALLOTROPIC TRANSFORMATION

The basis of allotropic transformation in context to crystalline solids has been explained in Chapter 1, while discussing the crystal structure of metals. Some of the metals such as iron, cobalt, and manganese undergo this allotropic transformation. Allotropic transformation is usually represented in a binary phase diagram in form of points on the vertical line indicating pure metal. The respective solid solution of these allotropes further may be represented in phase diagram in form of closed loop. Iron–carbon phase diagram is a nice example of such transformation, and the details of this will be explained in the next chapter.

5.5.2 ORDER–DISORDER TRANSFORMATION

The concept of ordered and random solid solutions has been explained in Chapter 4. In general, the solute atoms in a substitutional solid solution occupy random locations yielding disordered solid solutions. However, in certain cases, precise control over the cooling rate may result a particular arrangement of solute atoms in the alloy, known as ordered solid solution. In most of the cases, ordering takes place with a particular composition corresponding to certain atomic ratio. As a convention, the ordered phases in a phase diagram are usually denoted by α', β', γ' or α', α'', and so on. And the phase boundary of these order phases is drawn by dot–dash lines.

5.5.3 EUTECTOID REACTION

This is a most commonly occurring phase transformation in solid state, where a solid upon cooling produces two different solids as represented in the following. A very familiarized eutectoid reaction exists in iron–carbon diagram, where austenite converts to a mixture of ferrite and cementite (known as pearlite) and will be discussed in detail in the next chapter.

$$\text{Solid } \alpha \underset{\text{Heating}}{\overset{\text{Cooling}}{\rightleftharpoons}} \text{Solid } \beta + \text{Solid } \gamma$$

5.5.4 PERITECTOID REACTION

When a mixture of two different solids upon cooling yields a new solid, the transformation is known as peritectoid transformation as presented in the following:

$$\text{Solid } \alpha + \text{Solid } \beta \underset{\text{Heating}}{\overset{\text{Cooling}}{\rightleftharpoons}} \text{Solid } \gamma$$

Phase Diagrams and Phase Transformations

Usually, the new solid formed is a terminal solid solution or an intermediate phase. Ni–Zn and Cu–Si alloys are some of examples showing peritectoid reaction.

5.6 ROLES OF DEFECTS AND DIFFUSION

Diffusion has a very important role to play in the case of solid-state phase transformation. The movement of atoms in a solid body is usually random in nature. However, in case where a large number of such diffusing atoms are involved, the flow may trace a systematic path. As discussed in eutectic reaction, alternate layers of solids form the eutectic mixture. And this requires diffusion of the solute atoms to ensure appropriate concentration in each layer. The case is also similar in the case of eutectoid reaction.

Diffusion in solid state usually takes place by three mechanisms: (i) vacancy diffusion, (ii) interstitial diffusion, and (iii) atom exchange mechanism. As stated earlier, vacancy is a thermodynamic feature in a crystalline solid, and the presence of vacancy sites is usually preferred for ease diffusion. As shown in Figure 5.11, an atom sitting next to the vacancy site may jump the vacancy site, thus creating vacancy in its earlier location.

In a similar way, an interstitial atom may occupy a normal lattice site repelling an atom from its lattice site to a nearby interstitial location as shown in Figure 5.12.

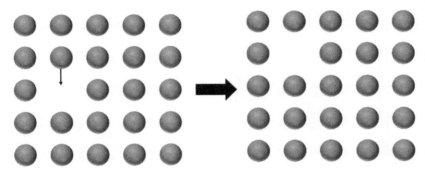

FIGURE 5.11 Schematic representation of vacancy diffusion in crystalline solid.

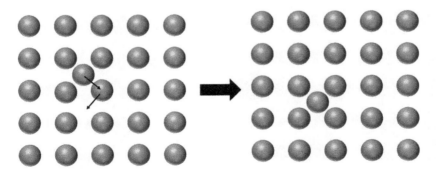

FIGURE 5.12 Schematic representation of interstitial diffusion in crystalline solids.

78 Phase Transformations and Heat Treatments of Steels

The other, i.e., atom mutual interchange making a close loop is also a possible diffusion mode. But various experimental evidences have shown vacancy diffusion to be the primary mode of diffusion in solids. The rate at which diffusion takes place is affected by several parameters, such as temperature, concentration, surface area, and so on.

FURTHER READING

Avner, S. H. *Introduction to Physical Metallurgy*. (Tata McGraw-Hill Education, New York, 1997).

Clark, D. S. & Varney, W. R. *Physical Metallurgy for Engineers*. (Litton Educational Publishing Inc., New York, 1962).

6 Iron–Carbon Phase Diagram

6.1 INTRODUCTION

Iron and carbon alloys, particularly steels, are widely used engineering alloys because these are relatively cheaper with an extensive range of properties. A study of the iron–carbon system is therefore valuable, as it helps us to understand the phase transformations that occur in steels. Before we consider steels, it is equally important to know the behavior of pure iron. The purest form of iron, which is commonly used in industries, is called as ingot iron. It has a typical tensile strength of about 300 N/mm^2 and is extensively used for drainage culverts, roofing, ducts, washing machines, and stoves. Later, wrought iron, which is a mixture of high-purity iron and slag, was manufactured by the Byers process. In this process, the iron metal is melted and purified to a highly refined state; an iron silicate slag is prepared; finally, the iron metal is disintegrated and mechanically incorporated in the slag to form sponge-like balls of iron globules coated with the slag. These balls are hammered to squeeze out the excess slag and then rolled to bars, plates, rods, and tubes. A comparison of the composition and typical properties of ingot and wrought iron is given in Table 6.1.

The characteristic properties of wrought iron are because of the nature of the slag distribution, i.e., the presence of threadlike slag fibers in the soft ferrite matrix, such as good machinability, excellent shock and corrosion resistance, and good weldability. Wrought iron has a wide range of industrial applications, including railroads, ship buildings, oil industries, crane hooks, anchors, and architectural purposes.

TABLE 6.1
Composition and Mechanical Properties of Different Forms of Iron

% Composition	Ingot Iron	Wrought Iron
Carbon	0.012	0.06
Manganese	0.017	0.045
Silicon	Trace	0.101
Phosphorous	0.005	0.068
Sulfur	0.025	0.009
Slag		1.97
Mechanical Properties	**Longitudinal**	
Tensile strength (N/mm^2)	300	350
Yield strength (N/mm^2)	200	215
% elongation	75	18–25
% reduction in area	30–40	35–45

6.2 ALLOTROPIC TRANSFORMATIONS IN IRON

Iron exists in more than one type of lattice structure, depending upon the temperature. To understand the same, let us look at the typical cooling curve for pure iron, which is shown in Figure 6.1. Iron first solidifies at 1539°C. The iron at this stage is in the body-centered cubic (BCC) structure and is called delta iron (δ-Fe). Upon more cooling, δ-Fe changes to gamma iron (γ-Fe) at 1394°C. γ-Fe is face-centered cubic (FCC) and nonmagnetic. Furthermore, at 910°C, γ-Fe transforms to alpha iron (α-Fe), which is nonmagnetic, but the structure is changed to BCC. Finally, the α-Fe becomes magnetic without a change in crystal structure at 768°C. This temperature is commonly known as the Curie temperature (named after Madame Curie).

6.3 SOLUBILITY OF CARBON IN IRON

Carbon, when dissolved in iron, forms an interstitial solid solution. The solubility of carbon depends upon the type and nature of the interstitial voids and the crystal structure of iron (BCC or FCC). In the case of FCC γ-Fe, carbon enters the octahedral voids because the radius of the largest interstitial sphere that would just fit in octahedral voids is 0.052 nm as against 0.028 nm in tetrahedral voids. Figure 6.2a illustrates the octahedral holes in FCC lattice, which are also symmetrical. However, the radius of the carbon atom being 0.077 nm, some distortion of the lattice occurs

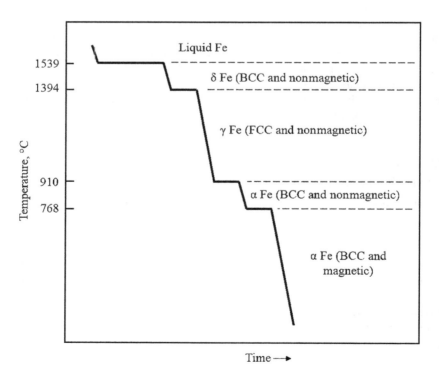

FIGURE 6.1 A typical cooling curve of pure iron.

Iron–Carbon Phase Diagram

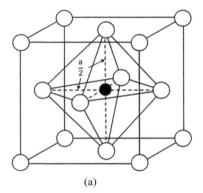
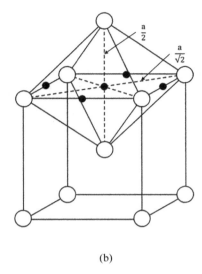

FIGURE 6.2 (a) Symmetrical octahedral void in FCC structure and (b) unsymmetrical octahedral void in BCC structure. The black spheres indicate the octahedral voids. It can also be seen that the size of the octahedral voids is larger in FCC than in BCC. BCC, body-centered cubic; FCC, face-centered cubic.

when the carbon atom enters the iron lattice. This lattice distortion limits the solubility of carbon in γ-Fe up to 2% at 1147°C and 0.8% at 727°C.

In the case of BCC α-Fe, the radius of the largest interstitial sphere that would just fit in tetrahedral voids (0.036 nm) is more than the octahedral voids (0.019 nm). Nevertheless, it is essential to note that the carbon atoms prefer to enter the smaller octahedral voids. In octahedral voids of BCC lattice, the carbon atom has only two nearest iron atoms, while four other iron atoms are at a larger distance, as shown in Figure 6.2b. Hence, the carbon atom displaces these two nearest atoms by 0.053 nm in one of the <100> directions to maintain symmetry (Figure 6.3a). It results in enlargement of the c-axis as compared with the a-axis, thereby leading to tetragonal distortion of the lattice. In contrast, when the carbon atom tends to occupy the larger tetrahedral voids, it has all the four atoms as its nearest neighbors, and displacing these four atoms causes more strain energy, as illustrated in Figure 6.3b. Therefore, the carbon atoms prefer to enter the smaller octahedral voids, as they cause less distortion than bigger tetrahedral voids.

It can also be seen in Figure 6.4 that the solubility of carbon in austenite is more than in ferrite, although the vacant space in the BCC lattice is greater than that in the FCC lattice. As discussed earlier, the presence of carbon in BCC α-Fe lattice causes tetragonal distortion, which restricts the iron to accommodate only a limited amount of carbon atoms. Hence, the maximum solubility of carbon in α-Fe is only 0.02% at 727°C. On comparison, the size of the octahedral voids in FCC γ-Fe lattice where carbon sits is larger, resulting in fewer distortions.

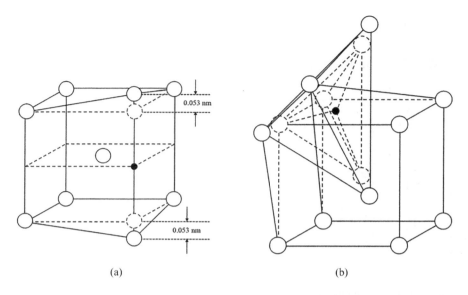

FIGURE 6.3 (a) Tetragonal distortion caused by the carbon atom (black sphere) in octahedral voids in BCC α-Fe and (b) carbon atom placed in the tetrahedral void in BCC α-Fe displaces more iron atoms resulting in more strains. BCC, body-centered cubic

6.4 IRON–IRON CARBIDE PHASE DIAGRAM

The iron–iron carbide phase diagram is only a part of the iron–carbon phase diagram, where the influence of carbon up to 6.67% by weight on the allotropic transformation of iron is studied. Iron forms an interstitial compound, iron carbide or cementite, Fe_3C, which contains 6.67% carbon. Therefore, this diagram is called the iron–iron carbide phase diagram. Many books prefer to call this diagram as an equilibrium diagram, but it is not an exact equilibrium diagram since equilibrium means no change of phase with time. Moreover, in this case, cementite is a metastable phase, and it decomposes to the stable form of carbon, i.e., graphite. This decomposition takes a very long time, and therefore, the metastable conditions are considered to represent the equilibrium changes.

6.4.1 Invariant Points in Fe–Fe$_3$C Phase Diagram

The iron–iron carbide phase diagram schematically illustrated in Figure 6.4 shows three horizontal lines where three important invariant reactions are described. The first invariant reaction, on cooling, occurs at 1495°C, which is represented as

$$L + \delta \rightleftharpoons \gamma \tag{6.1}$$

While gamma, γ, is an interstitial solid solution of carbon in γ-Fe, δ is an interstitial solid solution of carbon in δ-Fe. The maximum solubility of carbon in δ-Fe is only 0.10% (point A in Figure 6.5), whereas it is much more in γ-Fe. On increasing the

Iron–Carbon Phase Diagram

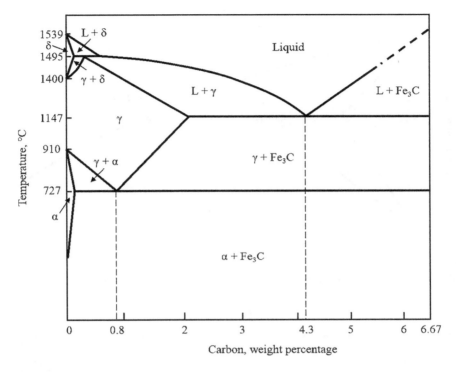

FIGURE 6.4 The iron–iron carbide phase diagram labeled in general terms.

amount of carbon up to 0.10%, the temperature of the allotropic change increases from 1400°C to 1495°C. The line AB represents the start of the δ to γ allotropic transformation for alloys containing less than 0.10% carbon. On the other hand, the horizontal line AC marks the beginning of the above transformation by the peritectic reaction (equation 6.1) for alloys containing carbon between 0.10% and 0.18%. The line BC represents the end of this δ to γ transformation for alloys containing less than 0.18% carbon. Similarly, line CD marks the start and end of the transformation through the peritectic reaction for alloys between 0.18% and 0.50% carbon.

The second horizontal line EFG in Figure 6.6 represents a eutectic reaction at 1147°C given by

$$L \rightleftharpoons \gamma + Fe_3C \tag{6.2}$$

Whenever the liquid crosses this line EFG at the eutectic point F, the eutectic reaction takes place, and the liquid solidifies into a mixture of two phases, namely, austenite (γ) and cementite (Fe_3C). This eutectic mixture is commonly known as ledeburite. It is usually not seen in microstructures since austenite undergoes another transformation during further cooling. In general, the alloys that are having more than 2.11% carbon are categorized under cast irons. Furthermore, the cast irons are subdivided into two classes. The alloys having carbon between 2.11% and 4.3% are

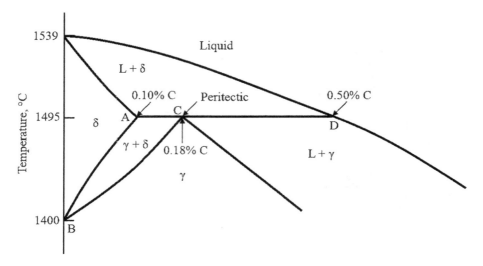

FIGURE 6.5 The peritectic region of the iron–iron carbide phase diagram.

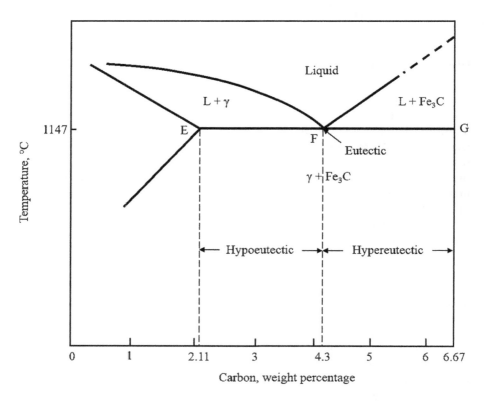

FIGURE 6.6 The eutectic region of the iron–iron carbide phase diagram.

Iron–Carbon Phase Diagram

called as hypoeutectic cast irons, whereas those having carbon between 4.3% and 6.67% are called as hypereutectic cast irons. Depending upon this classification, which is based on the amount of carbon, the composition of the alloy at the eutectic temperature varies. For example, the hypoeutectic cast irons (say 3% C) consist of proeutectic austenite along with liquid before the eutectic reaction takes place. In contrast, hypereutectic cast irons (say 5% C) comprise proeutectic cementite. The exact amount of these proeutectic phases is determined by using the Lever rule. The remaining amount of liquid transforms into the mixture of austenite and cementite, i.e., ledeburite upon cooling at the eutectic temperature.

The third horizontal line represents the eutectoid invariant reaction, and it is given by

$$\gamma \rightleftharpoons \alpha + Fe_3C \tag{6.3}$$

The eutectoid point is at 0.8% carbon, and the eutectoid temperature is 727°C (Figure 6.7). As described in equation 6.3, austenite transforms into a mixture of ferrite and cementite, commonly known as pearlite, because of its pearly appearance under the optical microscope. The detailed microstructure evolution of pearlite is discussed in Chapter 8. Like cast irons, steels that contain less than 2.11% carbon are

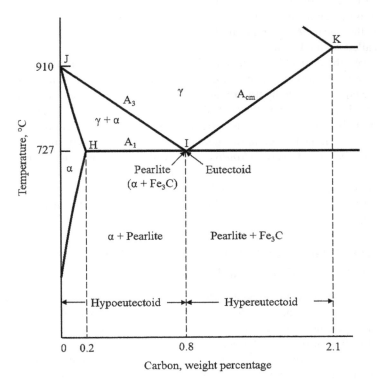

FIGURE 6.7 The eutectoid region of the iron–iron carbide phase diagram.

86 Phase Transformations and Heat Treatments of Steels

also subdivided into two different categories based on the carbon content. Steels that contain carbon between 0.02% and 0.8% are called hypoeutectoid steels, whereas hypereutectoid steels contain carbon between 0.8% and 2.11%. This steel portion of the iron–iron carbide phase diagram is of great interest because of the ample application areas of various types of steels. The properties of steels depend on their microstructures, and therefore, the microstructural changes that occur during cooling from the austenite range will be discussed in Chapter 11.

6.4.2 Critical Temperatures in Fe–Fe₃C Phase Diagram

In Figure 6.7, we have discussed the phase transformation upon slow cooling from austenite. The allotropic change from FCC γ-Fe to BCC α-Fe occurs when the hypoeutectoid steel crosses the IJ line. This line is known as the upper critical temperature and is often labeled in many books as A_3. Similarly, there are other critical temperature lines in the iron–iron carbide phase diagram, where phase transformations happen. For example, the eutectoid temperature line HI is called as the lower critical temperature line or the A_1 line, where austenite transforms to pearlite upon cooling. The line IK, which represents the maximum amount of carbon that can be dissolved in austenite of hypereutectoid steels as a function of temperature, is known as the upper critical temperature or A_{cm} line. When the hypereutectoid steels cross this A_{cm} line, the excess carbon above the amount required to saturate austenite precipitates as cementite along the grain boundaries of austenite. The Curie temperature line, i.e., the line at 768°C in Figure 6.8, where the α-Fe becomes magnetic on cooling, is called as the A_2 line.

However, the critical lines on cooling are not the same as for heating. The phase transformations always occur below the equilibrium temperature during cooling and above the equilibrium temperature during heating for which the critical line during cooling is always lower than the critical line during heating (Figure 6.8). The critical temperature line during heating is designated as A_c (c is taken from the French word *chauffage*, which means heating), whereas during cooling, it is designated as A_r (r is taken from the French word *refroidissement*, which means cooling). The difference between these two lines decreases with the decrease in heating rate. In other words, the two lines approach each other as the rate of heating and cooling becomes slower. Hence, at infinitely slow heating and cooling, the two lines appear precisely at the same temperature.

6.5 EFFECT OF ALLOYING ELEMENTS ON THE IRON–CARBON EQUILIBRIUM DIAGRAM

The presence of alloying elements affects the iron–carbon equilibrium diagram in such a way that the diagram no longer represents equilibrium conditions. The alloying elements change the eutectoid temperature, the position of the eutectoid point, and the location of the α- and γ-fields. The influence of different alloying elements on eutectoid temperature is shown in Figure 6.9. As evident from the figure, Mn and Ni tend to lower the eutectoid temperature. Other alloying elements such as Cr, V, W, and Si raise the eutectoid temperature, as their concentration increases. Ti and Mo

Iron–Carbon Phase Diagram

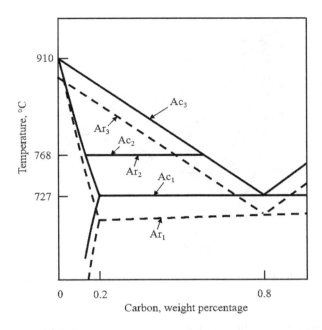

FIGURE 6.8 Hypoeutectoid region of the iron–iron carbide phase diagram showing various critical temperature lines.

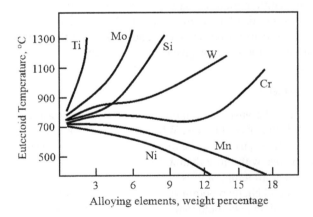

FIGURE 6.9 Influence of several alloying elements on the eutectoid temperature.

are most effective in increasing the eutectoid temperature. This shift in the eutectoid temperature is essential to note since it decides whether it will raise or lower the proper hardening temperature as compared to plain carbon steels.

In contrast, all the alloying elements tend to lower the carbon content of the eutectoid composition (Figure 6.10). Ti and Mo are the most effective in lowering the eutectoid composition. It may also be noted that increasing amounts of Mn and Ni

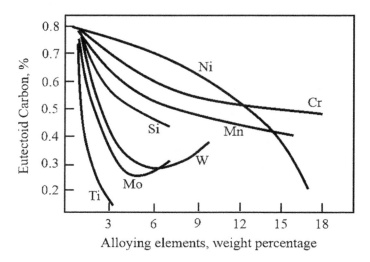

FIGURE 6.10 Influence of several alloying elements on the eutectoid composition.

lower the critical temperature sufficiently so that austenite does not transform even on slow cooling, and hence, it is retained at room temperature. These elements, along with Co, Cu, and Zn, are, therefore, categorized as austenite-stabilizing elements or austenite stabilizers. On the other hand, alloying elements such as Cr, W, Mo, V, and Si reduce the austenitic region and enlarge the field in which α- or δ-iron is found. Since they tend to stabilize ferrite, they are called ferrite stabilizers. These elements are more soluble in α-iron than in γ-iron and thus favor the formation of a larger quantity of carbide in the steel for given carbon content.

6.5.1 Open γ-Field

Mn and Ni enlarge the austenite region, make it a stable phase even at room temperature, and shift the critical points. While the A_4 point shifts upward, the A_3 point moves downward, thus narrowing the range of α-phase and enlarging that of γ-phase (Figure 6.11a). Alloys with more than $X\%$ of solute have a solid solution of alloying elements in γ-iron at all temperatures and, hence, are called austenitic alloys. These two elements, Mn and Ni, help obtain austenitic steels. For example, austenitic stainless steel with 8% Ni and Hadfield manganese steel with about 12% Mn have a wide range of industrial applications. Some other elements such as Rh, Pd, Cr, Ru, Co, Pt, Os, and Ir have similar effects.

6.5.2 Expanded γ-Field

C and N expand γ-region, but due to their limited solubility in iron and by the formation of a compound, the range of γ-field reduces, although the two-phase region containing γ-phase remains (Figure 6.11b).

Iron–Carbon Phase Diagram

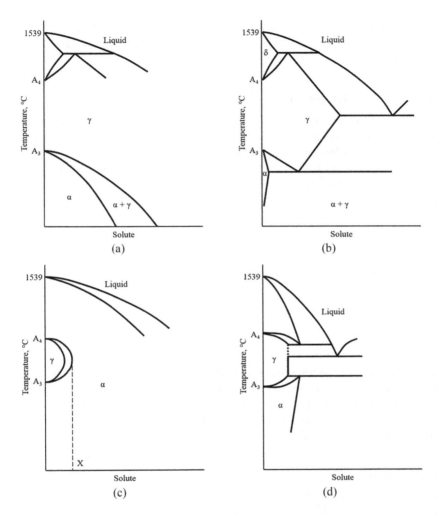

FIGURE 6.11 Different categories of the iron–iron carbide phase diagram: (a) open γ-field, (b) expanded γ-field, (c) closed γ-field, and (d) contracted γ-field.

6.5.3 Closed γ-Field

Elements such as Si, Al, Be, and P lower A_4 and raise A_3 to contract the γ-region to a small area called γ-loop, as shown in Figure 6.11c. Alloys with more than $X\%$ of solute in Figure 6.11c have a solid solution of alloying elements in γ-iron at all temperatures and, therefore, are called ferritic alloys. Influential carbide formers such as Cr, W, Mo, V, and Ti have similar effects.

The austenite phase may not appear in steel at any temperature when sufficiently large quantities of such elements are added. For example, by adding 12.8% Cr or more, the austenitic phase disappears, and the δ-ferrite and α-ferrite merge to give a

90   Phase Transformations and Heat Treatments of Steels

continuous ferrite from room temperature up to the melting point. Such steels cannot be heat-treated conventionally because the austenite phase is not available for phase transformation.

6.5.4 CONTRACTED γ-FIELD

B, Ta, Nb, and Zr strongly contract the γ-region, but due to their low solubility in iron and by the formation of a compound, two-phase alloys are formed before the γ-phase region is wholly enclosed (Figure 6.11d).

Zener and Andrews have developed a thermodynamic relationship to explain the formation of open γ-field and closed γ-field in phase diagrams. The relationship is as follows:

$$\frac{C_\alpha}{C_\gamma} = \beta e^{\Delta H / RT} \tag{6.4}$$

or,

$$\ln \frac{C_\alpha}{C_\gamma} = \frac{\Delta H}{RT} \tag{6.5}$$

where C_α and C_γ are the fractional concentrations of an alloying element in α-Fe and γ-Fe, respectively; β is a constant; ΔH is the enthalpy change, i.e., $H_\gamma - H_\alpha$; H_γ is the heat absorbed per unit of solute dissolving in γ-Fe; and H_α is the heat absorbed per unit of solute dissolving in α-Fe. The solute forms a closed γ-loop if the enthalpy change, ΔH is positive, and an open γ-loop if it is negative.

FURTHER READING

Abbaschian, R. & Reed-Hill, R. E. *Physical Metallurgy Principles.* (Cengage Learning, Massachusetts, 2008).

Raghavan, V. *Physical Metallurgy: Principles and Practice.* (PHI Learning Pvt. Ltd., New Delhi, 2006).

Avner, S. H. *Introduction to Physical Metallurgy.* (Tata McGraw-Hill Education, New York, 1997).

7 Thermodynamics and Kinetics of Solid-State Phase Transformation

7.1 INTRODUCTION

A phase transformation always needs a driving force and happens only when there is a net decrement in the free energy of the system. In this regard, the thermodynamics of the feasibility of a phase transformation is essential to understand to comprehend the presence of possible phases under certain combinations of temperature, pressure, and so on. Furthermore, the conversion is not abrupt. It takes place at a fixed or dynamic rate, which is again governed by a set of parameters. Hence, a phase transformation is divided into two parts, namely, nucleation and growth.

7.2 NUCLEATION

Most phase transformations start by the formation of a large number of small particles of the new phase(s), and these particles subsequently grow in size until the transformation is complete. The appearance of the particles can be termed as nucleation, while growth involves the increase in the size of these nuclei. There are two types of nucleation: homogenous and heterogeneous. In homogenous nucleation, the probability of nucleation at any given site is almost similar to that of any other place within the assembly, whereas, in heterogeneous nucleation, the likelihood of nucleation at specific preferred locations in the assembly is much higher than that at other locations. For example, in gases, heterogeneous nucleation can occur at container walls or at any other impurity particles (Figure 7.1).

7.2.1 HOMOGENOUS NUCLEATION

In the case of the solidification of a liquid phase, which is undercooled below the melting temperature, a solid nucleus forms leading to a Gibbs free energy change of ΔG, which can be written as

$$\Delta G = -V_s \Delta G_v + A_{sl}\gamma \tag{7.1}$$

where V_s is the volume of the solid sphere of radius r, ΔG_v is the free energy difference per unit volume between the solid and liquid phases, or the volume free energy, A_{sl} is the solid/liquid interfacial area, and γ is the solid/liquid interfacial free energy.

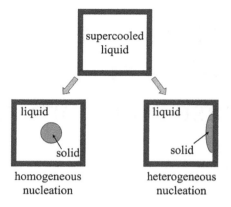

FIGURE 7.1 Two types of nucleation.

We can write ΔG_v as

$$\Delta G_v = G_{vl} - G_{vs} \tag{7.2}$$

where G_{vl} and G_{vs} are the free energies per unit volume of liquid and solid, respectively. It is apparent from Figure 7.2 that at $T < T_m$, $G_{vs} < G_{vl}$. Therefore, a driving force for solidification, ΔG_v, always exists whenever a liquid is cooled below the melting temperature.

At temperature T:

$$G_{vl} = H_{vl} - TS_{vl} \tag{7.3}$$

$$G_{vs} = H_{vs} - TS_{vs} \tag{7.4}$$

$$\Delta G_v = \Delta H_v - T\Delta S_v \tag{7.5}$$

At temperature T_m:

$$\Delta G_v = \Delta H_{mv} - T_m \Delta S_{mv} = 0 \tag{7.6}$$

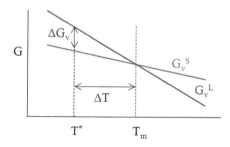

FIGURE 7.2 Variation of free energy of liquid and solid as a function of temperature.

Solid-State Phase Transformation

$$\Delta S_{mv} = \frac{\Delta H_{mv}}{T_m} \tag{7.7}$$

ΔH_v and ΔS_v can be considered as independent of temperature for small undercooling ΔT:

$$\Delta G_v \approx \Delta H_{mv} - \frac{T \Delta H_{mv}}{T_m} = \frac{\Delta H_{mv} \Delta T}{T_m} \tag{7.8}$$

where ΔH_{mv} is the latent heat of fusion per unit volume.

Below T_m, ΔG_v is positive. Hence, the free energy change due to the formation of a small volume of liquid has a negative input. However, γ is always positive due to the creation of a solid/liquid interface. Now, for the solid nucleus having radius r, we can have

$$\Delta G = -\frac{4}{3}\pi r^3 \Delta G_v + 4\pi r^2 \gamma \tag{7.9}$$

This function first rises, passes through a maximum, and finally decreases, as shown in Figure 7.3. In other words, the free energy increases when a solid particle begins to form. When the solid cluster reaches a size equivalent to the critical radius r^*, it grows with a decrease in free energy. On the other hand, a solid cluster of radius $r < r^*$ shrinks and redissolves. This particle is called an embryo, whereas the particle of radius $r > r^*$ is termed a nucleus. Hence, the critical or the maximum free energy, ΔG^*, corresponds to an activation free energy, which is the free energy required for the formation of a stable nucleus.

Setting $\dfrac{d \Delta G}{dr} = 0$, we have

$$r^* = -\frac{2\gamma}{\Delta G_v} \tag{7.10}$$

$$\Delta G^* = \frac{16 \pi \gamma^3}{3 (\Delta G_v)^2} \tag{7.11}$$

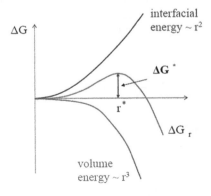

FIGURE 7.3 Free energy change of the nucleus as a function of its radius.

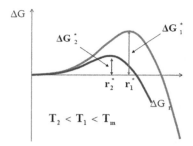

FIGURE 7.4 Critical radii and total free energy at different temperatures.

Substituting the value for ΔG_v, we have

$$r^* = \left(\frac{2\gamma T_m}{\Delta H_m}\right)\frac{1}{\Delta T} \tag{7.12}$$

$$\Delta G^* = \left(\frac{16\pi\gamma^3 (T_m)^2}{3(\Delta H_m)^2}\right)\frac{1}{(\Delta T)^2} \tag{7.13}$$

It can be noted that both r^* and ΔG^* decrease with increasing undercooling (ΔT), as shown in Figure 7.4.

7.2.1.1 Rate of Homogenous Nucleation

The nucleation process can be defined as the addition of one atom to a critical-sized particle to make it just supercritical. Let us suppose that there are s^* atoms in the liquid phase facing the critical-sized particle across the interface. If anyone of these atoms jumps from the liquid phase to the solid phase, the particle becomes just supercritical and is said to be nucleated. Growth continues if additional atoms are added to the supercritical particles.

The Maxwell–Boltzmann statistics can be used to know the number of critical-sized particles in the liquid phase. Let us suppose that the total number of particles per unit volume of the liquid phase be N_T. The number of critical-sized particles N^* is approximated by

$$N^* = N_T \exp\left(-\frac{\Delta G^*}{KT}\right) \tag{7.14}$$

where G^* is the nucleation barrier. The frequency with which the s^* atoms neighboring the critical-sized particle can cross the interface to join the particle is given by

$$v' = s^* v \exp\left(-\frac{\Delta G_D}{KT}\right) \tag{7.15}$$

Solid-State Phase Transformation

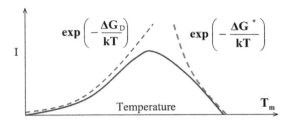

FIGURE 7.5 The homogenous nucleation rate as a function of temperature.

where ν is the lattice vibration frequency (which is ~10^{13}/s above the Debye temperature) and ΔG_D is the activation barrier for diffusion across the interface. Now, the rate of nucleation $I\left(=\dfrac{dN}{dt}\right)$ is given by

$$I = N \times \nu' = N_T \nu s^* \exp\left[-\dfrac{\left(\Delta G^* + \Delta G_D\right)}{KT}\right] \tag{7.16}$$

At the temperature of equilibrium between two phases, T_0, when undercooling is zero ($T = 0$), ΔG^* is infinite. Therefore, the nucleation rate I, as given by equation 7.16, is zero at the equilibrium temperature. As the temperature decreases, $\Delta T > 0$ and ΔG becomes finite. As ΔG decreases with decreasing temperature, I increases. But as the temperature decreases, this at the same time diminishes the probability of attachment of atoms to a critical nucleus. Therefore, as the term $\exp\left(-\dfrac{\Delta G_D}{KT}\right)$ decreases rapidly, the nucleation rate decreases with decreasing temperature. It becomes zero at 0 K. Because of these two opposing factors, I reaches a maximum value, as shown in Figure 7.5.

The temperature of the maximum rate of nucleation $T_{I_{\max}}$, can be found by setting $\dfrac{dI}{dt} = 0$, from which the following expression for $T_{I_{\max}}$ can be obtained.

$$\dfrac{d\Delta G^*}{dT} = \dfrac{\left(\Delta G^* + \Delta G_D\right)}{T_{I_{\max}}} \tag{7.17}$$

7.2.2 Heterogeneous Nucleation

Nucleation in the solid state, as in the liquid state, is often heterogeneous or nonrandom. Preferred nucleation sites include grain boundaries, grain corners, grain edges, dislocations, and surfaces of inclusions or precipitates embedded in the matrix, as shown in Figure 7.6.

In the case of heterogeneous nucleation of a spherical cap on a wall of a container (Figure 7.7), three interfacial energies γ_{lc} – liquid–container interface, γ_{ls} – liquid–solid interface, and γ_{sc} – solid–container interface can be considered.

FIGURE 7.6 Schematic of heterogeneous nucleation.

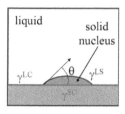

FIGURE 7.7 Quasi-equilibrium between three interfacial energies associated with the solid nucleus.

By balancing the interfacial forces in the plane of the container wall, we have

$$\gamma_{lc} = \gamma_{sc} + \gamma_{ls} \cos\theta \tag{7.18}$$

$$\cos\theta = (\gamma_{lc} - \gamma_{sc})/\gamma_{ls} \tag{7.19}$$

where θ is the wetting angle.

The creation of the nucleus leads to a Gibbs free energy change ΔG_{het}. Hence,

$$\Delta G_{het} = -V_s \Delta G_v + A_{sl}\gamma_{sl} + A_{sc}\gamma_{sc} - A_{sc}\gamma_{lc} \tag{7.20}$$

Here,

$$V_s = \pi r^3 (2 + \cos\theta)(1 + \cos\theta)^2 / 3 \tag{7.21}$$

$$A_{sl} = 2\pi r^2 (1 - \cos\theta) \tag{7.22}$$

$$A_{sc} = \pi r^2 \sin^2\theta \tag{7.23}$$

Therefore,

$$\Delta G_{het} = \left\{ -\frac{4}{3}\pi r^3 \Delta G_v + 4\pi r^2 \gamma_{sl} \right\} S(\theta) \tag{7.24}$$

$$\Delta G_{het} = \Delta G_{hom} S(\theta) \tag{7.25}$$

where $S(\theta)$ is a shape factor given by

$$S(\theta) = (2 + \cos\theta)(1 - \cos\theta)^2 / 4 \tag{7.26}$$

Solid-State Phase Transformation

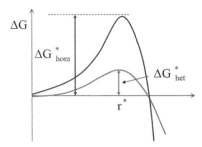

FIGURE 7.8 The excess free energy of solid for different types of nucleation. Note that r^* is independent of the nucleation site.

At $r = r^*$,

$$\frac{d\Delta G_{het}}{dr} = \left(-4\pi r^2 \Delta G_v + 8\pi r \gamma_{sl}\right) S(\theta) = 0 \tag{7.27}$$

$$r^* = \frac{2\gamma_{sl}}{\Delta G_v} \tag{7.28}$$

$$\Delta G^*_{het} = \frac{16\pi (\gamma_{sl})^3}{3(\Delta G_v)^2} S(\theta) \tag{7.29}$$

$$\Delta G^*_{het} = \Delta G^*_{hom} S(\theta) \tag{7.30}$$

$S(\theta)$ has a numerical value ≤1 dependent only on θ, i.e., the shape of the nucleus. Hence, it is often referred to as a shape factor. Two specials cases need consideration:

Case I: $\theta \to 90°$, $\cos\theta \to 0$, $\Delta G^*_{het} \to \frac{1}{2}\Delta G^*_{hom}$.

The hemispherical-shaped solid is still effective since the energy barrier is half of that of the homogenous nucleation.

Case II: $\theta \to 0°$, $\cos\theta \to 1$, $\Delta G^*_{het} \to 0$.

Since there is no energy barrier to the heterogeneous nucleation, it can start just at the freezing temperature.

It can be concluded that when the contact angle is small, heterogeneous nucleation becomes easier. The barrier energy required for heterogeneous nucleation decreases, as shown in Figure 7.8.

7.2.2.1 Rate of Heterogeneous Nucleation

The rate of heterogeneous nucleation can be expressed on analogous arguments for homogenous nucleation. The number of nucleation sites can be given by

$$N^*_{het} = N_T \exp\left(-\frac{\Delta G^*_{het}}{KT}\right) \tag{7.31}$$

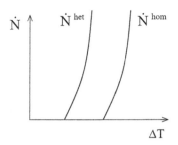

FIGURE 7.9 Variation of homogenous and heterogeneous nucleation rates with undercooling assuming the same critical value of ΔG^*.

$$N^*_{\text{hom}} = N_T \exp\left(-\frac{\Delta G^*_{\text{hom}}}{KT}\right) \quad (7.32)$$

$$N^*_{\text{het}} \gg N^*_{\text{hom}} \quad (7.33)$$

It is clear from Figure 7.9 that the heterogeneous nucleation starts at a lower undercooling. Now, the rate of homogenous nucleation can be written as

$$I_{\text{het}} = N_T \nu s^* \exp\left[-\frac{\left(\Delta G^*_{\text{het}} + \Delta G_D\right)}{KT}\right] \quad (7.34)$$

Thus, $I_{\text{het}} > I_{\text{hom}}$, because $\Delta G^*_{\text{het}} < \Delta G^*_{\text{hom}}$.

7.3 GROWTH KINETICS

Once the solid nucleus has exceeded the critical size and becomes a stable nucleus, growth occurs. In line with nucleation kinetics, there are also many facets to study the growth kinetics. In this context, we shall discuss two types of growth: interface-controlled growth and diffusion-controlled growth. In the case of interface-controlled growth, a nucleated particle can grow spontaneously by adding atoms or molecules at its surface or interface. One such example is the freezing of water that involves the movement of water molecules across the liquid/solid interface. The movement needs thermal activation and is short range.

In contrast, diffusion-controlled growth requires long-range diffusion. For example, during precipitation from a supersaturated solution, the critical nucleus gets enriched with one of the components in relation to the average composition of the neighboring matrix.

7.3.1 Interface-Controlled Growth

We shall assume here that the interface is nonsingular and incoherent. The precipitate differs in composition from the matrix. However, the rate of growth is controlled by the mechanism that allows the solute atoms to cross over from the matrix to

Solid-State Phase Transformation

the precipitate. Let ΔG_D be the activation energy for an atomic jump across the interface. The change in the chemical free energy per atom in the $\alpha \to \beta$ transformation is $v\Delta g$, where v is the volume per atom. The free energy barrier for an atomic jump from α to β is ΔG_D, while the jump from β back to α is $\Delta G_D - v\Delta g$, as shown in Figure 7.10.

The net rate of atom transfer from α to β is equal to the difference in the forward and backward rate. Therefore, the net rate of atomic jumping from α to β per unit area of the interface is

$$I = s\nu \exp\left(-\frac{\Delta G_D}{kT}\right)\left[1 - \exp\left(\frac{v\Delta g}{kT}\right)\right] \tag{7.35}$$

where s is the number of interfacial atoms in each phase per unit area of interface and ν is the atomic vibration frequency.

Let us suppose that λ is the jump distance across the interface. The velocity of the boundary will, therefore, be given by

$$u = \frac{\lambda I}{s} \tag{7.36}$$

$$u = \lambda \nu \exp\left(-\frac{\Delta G_D}{kT}\right)\left[1 - \exp\left(\frac{v\Delta g}{kT}\right)\right] \tag{7.37}$$

For a sufficiently small undercooling, we may assume $v\Delta g \ll kT$. Hence, the growth velocity becomes

$$u = \lambda \nu \exp\left(-\frac{\Delta G_D}{kT}\right) \tag{7.38}$$

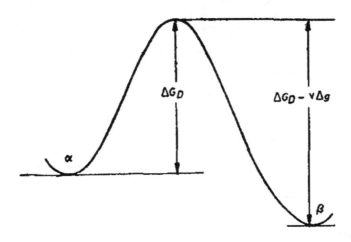

FIGURE 7.10 The free energy barrier for atom jump across the interface during growth.[1]

The diffusion coefficient D_b for the diffusional jump across the interface can be defined as

$$D_b = \lambda^2 v \, \exp\left(-\frac{\Delta G_D}{kT}\right) \qquad (7.39)$$

Expression (7.39) for the growth velocity shows that as T becomes very small, it approaches zero. Again, it is zero at the transformation temperature, T_0. It means that the growth velocity reaches a maximum value at some intermediate temperature. This is verified experimentally for the transformation of white to gray tin ($T_0 = +13°C$). The maximum in the growth rate (~1 mm/h) occurs at $-32°C$, where $\Delta T = 45°C$. After that, the growth rate decreases exponentially with temperature, as the boundary diffusion D_b, falls, as given by equation 7.39.

7.3.2 Diffusion-Controlled Growth

It refers to the formation of a new phase with a composition different from that of the old phase, and the rate is controlled by the long-range diffusion of one of the component atoms through the matrix phase. To understand this phenomenon, let us take a simple example of precipitation of β-phase particles from a supersaturated solid solution of α in a binary system of A and B. Zener proposed a simple theory for this type of growth. Let us consider Zener's theory for nonspherical precipitates, which grow as a plate in the direction normal to its surface. Figure 7.11 shows the precipitate particles of the β-phase formed in the matrix. The particles being abundant in B, the composition of the matrix changes toward the equilibrium value $c_{\alpha\beta} (< \bar{c})$.

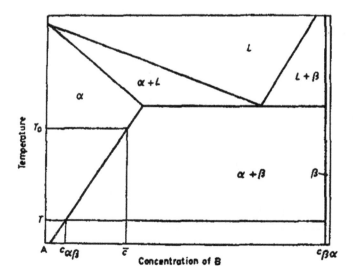

FIGURE 7.11 A schematic binary phase diagram, showing the decreasing solid solubility of B in A, as the temperature decreases.[1]

Solid-State Phase Transformation

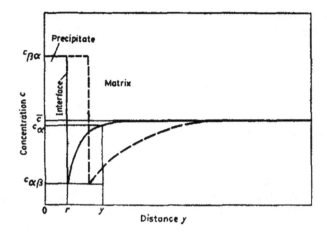

FIGURE 7.12 The concentration of B atoms as a function of distance during continuous precipitation of β.[1]

The concentration–distance profile during the precipitation is shown in Figure 7.12 along the distance axis, at $y = 0$. The precipitation starts in the matrix of the initial composition $\bar{c}$. The precipitate has grown to a size r, where the interface separating the precipitate from the matrix is located. The composition within the precipitate particle from $y = 0$ to $y = r$ is uniform and is equal to $c_{\beta\alpha}$. Equilibrium is assumed to prevail at the interface so that the concentration of B in the α-phase can be shown as $c_{\alpha\beta}$. Due to the jumping of atoms from the matrix into the precipitate, there is a drop in the concentration of B atoms in the matrix, as one moves away from the interface and reaches $\bar{c}$. As the precipitate particle grows to larger sizes, the region of the surrounding matrix depleted of B atoms extends to larger distances, as shown by the dotted line profile in Figure 7.12. c_α denotes the concentration of B atoms in the depleted region.

Let us suppose that in small time t, the boundary of the precipitate was moved forward into the matrix through a distance dx. As the composition at the boundary or interface remains independent of time, the number of solute atoms that arrive at the interface from the matrix must be equal to the number of solute atoms that are added to the growing particle. Hence, we have

$$\left(C_{\beta\alpha} - C_{\alpha\beta}\right) A dx = AD\left(\frac{dC_\alpha}{dx}\right) dt \qquad (7.40)$$

where D is the diffusion coefficient which is assumed to be independent of concentration and $\frac{dC_\alpha}{dx}$ is the concentration gradient of B in the matrix phase at the interface. Now, the growth rate of the particle is given by

$$u = \frac{dx}{dt} = \frac{D}{C_{\beta\alpha} - C_{\alpha\beta}} \frac{dC_\alpha}{dx} \qquad (7.41)$$

7.4 TIME–TEMPERATURE TRANSFORMATION AND CONTINUOUS COOLING TRANSFORMATION DIAGRAMS

The iron–iron carbide equilibrium phase diagram has very little significance in practice as it does not include the phase transformation kinetics into account. As most of the heat treatment practices are governed by nonequilibrium or fast cooling, the finished or final product carries the effect of the cooling rate and hence exhibits different properties. The equilibrium phase diagram also does not include the metastable phases like bainite and martensite. Thus, the requirement of an understanding of the correlation between temperature and time on the phase transformation is realized.

The first idea regarding the phase transformation kinetics was established by Davenport and Bain (Trans. AIME, vol. 90, p. 117, 1930). The diagram that reveals the phase transformation of austenite into various phases at a certain combination of temperature and time is known as a TTT curve. The TTT curve is based upon isothermal cooling of the steel specimen. Each composition of steel has its characteristic TTT curve. Although grain size and microstructural inclusions have some role to play in this context, in general, their effects are neglected.

Steel with the eutectoid composition of about 0.8 wt% carbon is chosen for a more straightforward interpretation of the TTT curve, as this does not contain any proeutectoid phase in it. Several thin pieces (as small pieces will respond to the temperature soon) of this composition are taken. All these specimens are heated to a sufficiently high temperature (~775°C) to get a completely austenitic structure in the furnace or molten bath. Another furnace or molten bath is kept ready at a specific subcritical temperature, i.e., below the lower critical temperature (~675°C). After completion of austenitization in the earlier furnace, all the specimens were transferred immediately to the later furnace. In this furnace, each sample is kept for different periods (e.g., 30 s, 5 h, 15 h, 25 h, 50 h, and 75 h) and then quenched in iced brine. The microstructural characterization of all the samples is done to analyze the overall pearlite fraction. A suitable way of representing both the time and temperature dependence of this transformation is shown in Figure 7.13.

For the sample, which was kept for very little time at the subcritical temperature, there is little chance of formation of pearlite, as it involves diffusional transformation, which requires enough time. Hence, quenching at this stage results in a martensitic structure. Increasing the time of stay at the subcritical temperature yields a higher pearlite content and lower martensite upon quenching; finally, beyond some time, the structure becomes pearlite entirely. Figure 7.14 represents the variation of pearlite percent in the final product with time at the subcritical temperature. The same methodology is followed for the same composition specimen but at different subcritical temperatures. It is exciting to know that at any particular temperature, the transformation rate is inversely proportional to the time required to complete the transformation to 100% pearlite. From Figure 7.14, it can be seen that at a temperature of 675°C, more time is required to complete the transformation. The transformation rate increases with decreasing temperature such that at 550°C, only about 10 s is required for 100% pearlite transformation. Thus, the shorter the time, the higher is the transformation rate.

Solid-State Phase Transformation

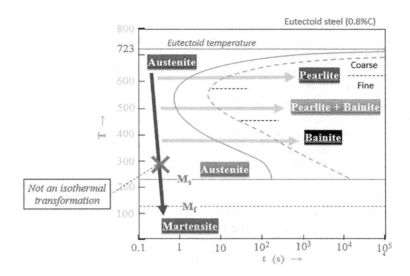

FIGURE 7.13 A typical TTT diagram for eutectoid steel. TTT, time–temperature transformation.

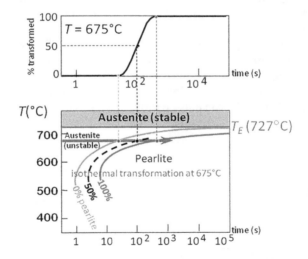

FIGURE 7.14 Variation of pearlite percent with the time stay at the subcritical temperature.

TTT diagrams are only strictly applicable to transformations carried out at constant temperatures. However, in industrial practices, steel is rarely quenched to a constant temperature. Instead, it is heated into the austenite range and continuously cooled to room temperature at different cooling rates. The relationships between transformation temperature and time can easily be depicted from continuous cooling transformation (CCT) diagrams. For simplicity, let us choose the steel of eutectoid composition. Figure 7.15 shows two cooling curves at different cooling rates to understand the CCT diagram. The specimens are cooled from above the eutectoid temperature to room temperature. As seen from Figure 7.15, curve 1 crosses the TTT diagram at point a. This point indicates the time required to nucleate pearlite isothermally at the temperature of point a, i.e., around 650°C.

At temperatures above 650°C, more time is required to start the pearlite transformation. Therefore, the point in continuous cooling where actually conversion starts lies to the right and below point a (designated as b).

Similarly, the finish of the pearlitic transformation can be shown as point d, whereas point c indicates the finish of the isothermal transformation. So, it can be concluded that the CCT lines are shifted to right with respect to the corresponding isothermal transformation lines. This also explains why there is no bainitic transformation in the metal in case of continuous cooling. The fact is that the pearlite range overshadows the bainitic range. In other words, austenite is converted to pearlite before the cooling curve reaches the bainitic transformation range. This happens in case of slow or moderate cooling rates (curve 1). Alternatively, in case of high heating rates (curve 2), the specimen stays in the bainitic transformation range for a shorter time to allow any significant bainite to form. Hence, in eutectoid steel, continuous cooling does not yield bainite.

It is clear that pearlite has its transformation start temperature (Ts) and transformation finish temperature (Tf) and, for bainite and martensite, the Ts (i.e., Bs and Ms) is independent of the cooling rate. For simplicity, we have marked Ts as heavy lines in Figure 7.16.

With decreasing cooling time or increasing cooling rate, Ts gradually decreases from the A_3 temperature to the Bs temperature. When a higher cooling rate is reached, the Ts jumps to the martensitic plateau. These plateaux signify the formation of metastable products, and they indicate that each metastable product possesses its Ts (Ms, Bs, and Ma), and it can only be formed within a certain cooling rate range.

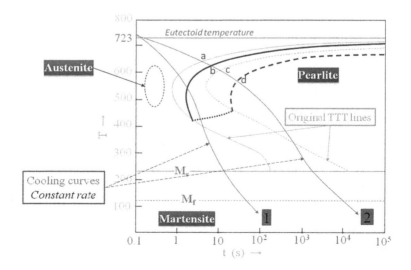

FIGURE 7.15 A typical continuous cooling transformation diagram.

Solid-State Phase Transformation

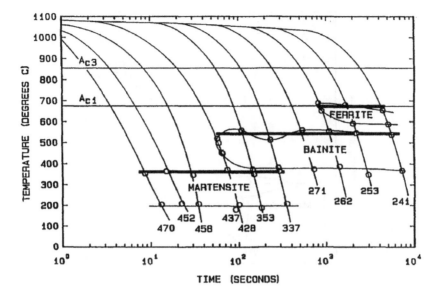

FIGURE 7.16 CCT diagram of 0.24C-1.67Mn-0.39Si-0.14Ni-0.17Cr-0.22Mo-0.11V.[2] CCT, continuous cooling transformation.

There also exists a minimum rate of quenching (critical cooling rate, CCR) that will produce a martensitic structure. This CCR misses the nose at which pearlitic transformation begins (Figure 7.17).

Recent advancements have proved that this is a misleading diagram as it shows that continuous cooling has to do something with the S-curve in TTT diagrams,

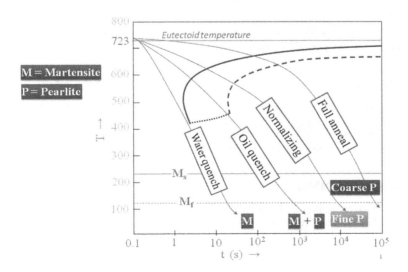

FIGURE 7.17 Effect of cooling rates on transformation.

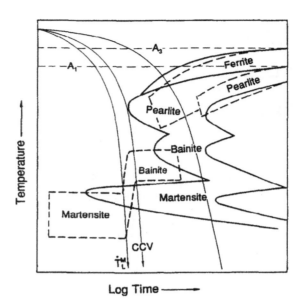

FIGURE 7.18 TTT diagram superimposed on a CCT diagram (dashed lines) to emphasize the difference in kinetics between the cooling transformations and isothermal transformations.[2] TTT, time–temperature transformation; CCT, continuous cooling transformation.

which is erroneous. The TTT curve is superimposed over the CCT diagram to understand the real mechanism, as shown in Figure 7.18.

It is not the CCR but the lower CCR limit of the martensite $\left(\dot{T}_L^M\right)$ that decides the formation of martensite during cooling. Depending upon the relative position and shape of the TTT and CCT diagrams, the CCR can be above or below the $\dot{T}_L^M$. In this case, CCR is lower than $\dot{T}_L^M$. Therefore, to analyze a complicated heat treatment procedure involving both cooling and isothermal transformations, CCT and TTT diagrams should be used together.[2]

REFERENCES

1. Sharma, R. C. Phase Transformations in Materials (PB). (CBS Publ., New Delhi, 2002).
2. Zhao, J.-C. & Notis, M. R. Continuous cooling transformation kinetics versus isothermal transformation kinetics of steels: A phenomenological rationalization of experimental observations. *Mater. Sci. Eng. R Rep.* **15**, 135–207 (1995).

FURTHER READING

Porter, D. A. & Easterling, K. E. *Phase Transformations in Metals and Alloys*, Third Edition (Revised Reprint). (CRC Press, Boca Raton, FL, 1992).

8 Phase Transformation in Steels

8.1 INTRODUCTION

The properties of steel are mostly dependent on the transformed product(s) from austenite, which are obtained during different heat treatment practices, and therefore, it is essential to understand the phase transformation in steels. Not only the presence of phases like pearlite, bainite, and martensite affect the properties, but also their morphology is equally significant in deciding the resultant properties. Hence, this chapter deals with the formation of austenite and its subsequent transformation behavior. Indeed, this study is essential in order to study and understand the theoretical aspects of available standard and modified heat treatment methods, which is described in detail in the subsequent chapter.

8.2 FORMATION OF AUSTENITE

The carbon steel contains a mixture of ferrite and pearlite, only pearlite, or pearlite and cementite at the room temperature depending on the carbon content of the steel. When the steel is heated into the austenite range, the formation of austenite occurs. For example, for hypoeutectoid steels, the formation of austenite starts when the temperature of the steel crosses the lower critical temperature, AC_1. Similarly, austenite starts to form just at the eutectoid temperature, i.e., 723°C for eutectoid steels. The transformation begins with the nucleation of austenite at the interfaces between the ferrite and cementite of pearlite, as shown in site 1 of Figure 8.1.

However, the more preferred site for nucleation is the boundary of the pearlite colony (site 2). This transformation process continues until all the ferrite and cementite dissolve to form austenite. While ferrite transforms to austenite by changing the crystal structure, cementite transforms to austenite by diffusion of carbon atoms. However, the process of dissolution of ferrite is faster and, therefore, gets completed before that of cementite. This results in the formation of inhomogeneous austenite.

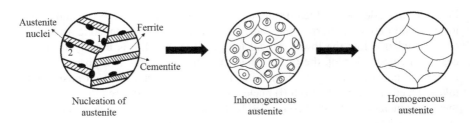

FIGURE 8.1 Different stages of austenite formation from lamellar pearlite.

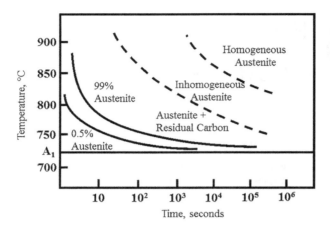

FIGURE 8.2 Effect of temperature and time on the austenite formation from pearlite in eutectoid steel.

Homogenization of austenite happens either by increasing the temperature or by holding the steel for a longer time so that all the carbon atoms diffuse. The complete transformation process is illustrated in Figure 8.2.

The grain size of austenite decides the properties of steel. So, different properties such as tensile strength, toughness, machinability, and hardenability can be changed by varying the austenite grain size. Based on the tendency of austenite grain to grain growth, there are two types of steels, namely, inherently fine-grained steels and inherently coarse-grained steels. Inherently fine-grained steels resist the grain growth of austenite with increasing temperature, whereas inherently coarse-grained steels grow abruptly with rising temperature (1000°C–1050°C), and hence, steels with coarse grains are obtained. However, once the temperature increases beyond 1050°C, the grain coarsening is quite rapid in steels with finer grain size, and at around 1150°C temperature, both the steel types have the same grain size (Figure 8.3). Since the inherently fine-grained steels are able to retain a fine austenite grain size for more extended periods, these are often used for the carburizing process at around 950°C with long carburizing cycles.

8.3 PEARLITIC TRANSFORMATION

Pearlite can be defined as the mixture of two distinct phases, i.e., ferrite and cementite in a well-defined pattern. When austenite having composition same as eutectoid one is cooled below the lower critical temperature, pearlite is usually formed. In case of a eutectoid steel, the active nuclei may be any of the two phases-ferrite or cementite, but usually it is the cementite. The rate at which pearlitic transformation takes place is governed by both (i) nucleation rate (N) and (ii) growth rate (G). Both these parameters are further controlled by the temperature at which transformation takes place. Nucleation rate is usually defined by the number of fresh nuclei that are formed per unit volume in a unit time in the untransformed austenite. It becomes

Phase Transformation in Steels

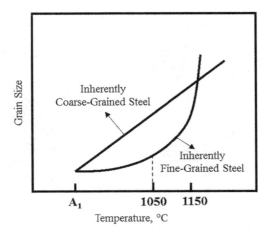

FIGURE 8.3 Effect of temperature on the austenite grain size in inherently fine-grained and coarse-grained steels.

higher as the temperature of transformation is lowered and at around the temperature close to 555°C, it apparently becomes maximum. At a constant temperature, the pearlitic growth is not a function of time. As temperature decreases below the eutectoid temperature, it increases and eventually reaches a maximum value close to the nose of the "S" curve. With further decrement in temperature, the rate again drops down. The morphology of pearlite is dependent on the relative rate of nucleation and growth, i.e. *N/G*. If this ratio is high, large numbers of nodules form but grow only a small distance as impingement occurs. With the fall of temperature, *N* increases at a faster rate than *G*, resulting in several pearlite nodules in one grain of austenite.

These pearlitic colonies tend to grow almost equally in both longitudinal and transverse directions of the lamella, resulting in a near spherical nodule. The most famous and widely accepted physical model for the formation of pearlite from austenite was proposed by F.C. Hull and R.F. Mehl and hence popularized as the Hull–Mehl model. Figure 8.4 shows different stages involved in the formation of pearlite from austenite. At first, the cementite nucleates at the austenite grain boundary and then grows. This growth occurs only by the removal of carbon from austenite, and consequently, austenite transforms to ferrite. Ferrite thus recently evolved starts growing along the cementite plate surface and thereby enriches the adjacent austenite with carbon. This process of nearby austenite enrichment by carbon continues till the concentration next to it approaches as that of cementite, and thus, a new cementite is generated next to the ferrite plate. In this process, other plates of ferrite and cementite are formed, and they form a colony.

8.3.1 Interlamellar Spacing

The interlamellar spacing of pearlite can be measured as the distance from the center of one ferrite (or may be taken as cementite) plate to immediate next ferrite (or cementite) plate in one pearlite colony. It is measured when the accounted

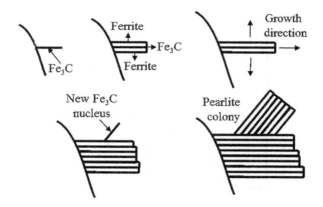

FIGURE 8.4 Different stages involved in the nucleation and growth of pearlite colony from austenite.

plates are perpendicular to the plane of the sample on which imaging is carried out. Assuming all other parameters to be the same, the interlamellar spacing is a function of the transformation temperature. At relatively higher transformation temperature, the interlamellar spacing is higher giving rise to formation of coarse pearlite. However, a finer pearlitic structure (i.e., smaller interlamellar spacing) usually possesses a higher strength. Alloying elements, except cobalt, increase the interlamellar spacing because they transform at higher temperatures. This process is not dependent on the prior grain size of austenite and the extent of concentration and structural homogeneity of austenite and hence can be termed as structure intensive. It decreases with decreasing transformation temperature; thus, pearlite becomes fine and finer.

8.3.2 Kinetics of Pearlite Transformation

Johnson and Mehl related the fraction of austenite transformed to pearlite as a function of time by the following equation:

$$f(t) = 1 - e^{(-\pi/3)} NG^3 + t^4 \tag{8.1}$$

Here, $f(t)$ is the fraction of austenite converted to pearlite. This equation has been developed based upon the following assumptions:

- N and G are constant with time.
- Nucleation is considered as a random process.
- The shape of the nodules remains spherical as they grow bigger (until they impinge on each other).

When the fraction transformed $f(t)$ is plotted against $\sqrt[4]{NG^3 t}$, a sigmoidal curve appears (Figure 8.5), which indicates that the kinetic behavior of pearlite formation is a nucleation and growth process.

Phase Transformation in Steels

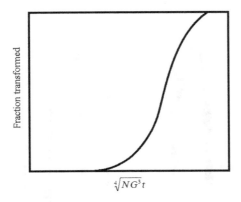

FIGURE 8.5 Reaction curve for nucleation.

8.4 BAINITIC TRANSFORMATION

Bainite is composed of ferrite and carbide with ferrite as the nucleus. However, unlike pearlite, bainite is not having a lamellar morphology, and cementite is distributed over the ferrite in a much finer scale. In case of a eutectoid steel, formation of bainite takes place at a temperature range of 200°C–500°C from undercooled austenite. In steels having disconnected pearlite and bainitic zones (in case of some alloy steels), bainitic transformation is possible by both isothermal and continuous cooling from austenite, whereas in the case of plain carbon steels and some other alloyed steels, overlapping of these pearlitic zones with the bainitic zone makes bainitic transformation not so easy. Depending on the cooling rate, either pearlite or bainite or a mixture may be formed. In case of very high rate of cooling, martensitic transformation may also be possible.

In such kind of steels, 100% bainitic steel may not be produced by continuous cooling. Rather, a faster cooling rate should be adopted in order to suppress pearlitic transformation and to make the cooling curve intersect the continuous cooling transformation curve in the bainitic region.

8.4.1 Upper Bainite

When the transformation temperature remains in the upper region of the bainitic zone, i.e., usually 550°C–400°C, upper bainite is formed. The morphology of upper bainite comprises lath-shaped ferrite, and the carbides that are precipitated out from the austenite matrix are nucleated along the needle axis of the bainite as shown in Figure 8.6. Upper bainite is also called feathery bainite due to its resemblance with the feather of a bird.

8.4.1.1 Formation of Upper Bainite

For formation of the upper bainite, ferrite is the active nucleus, and thus, there is a coherency between ferrite and the prior austenite. After nucleation of this ferrite, further growth takes place when the transformation is allowed to take place in

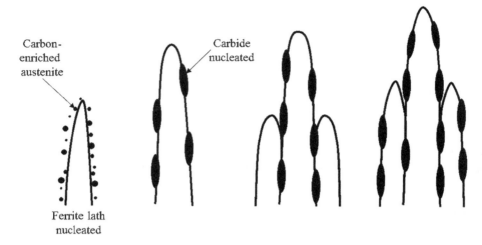

FIGURE 8.6 Different stages involved in the formation of upper bainite.

the upper bainite zone of the bainitic region. Due to rejection of carbon from this structure, the nearby austenite becomes eventually carbon rich and makes conditions favorable for cementite formation at the interface between the bainite and austenite. Growth of this generated carbide results in a carbon-depleted austenite next to it, again making conditions favorable for ferrite by the sheer mechanism. This process continues, and the growth of already-formed bainite and nucleation of fresh bainite take place simultaneously. Hereby, it can be inferred that diffusion and distribution of carbon in the austenitic region next to the bainitic structure is the dominant rate-controlling step in this case. It has also been postulated that various mechanisms of surface relief are also associated with the formation of upper bainite.

8.4.2 Lower Bainite

From the nomenclature, it is obvious that formation of the lower bainite is facilitated when the transformation is allowed to take place in the lower zone of the bainitic region, i.e., 250°C–400°C. Interestingly, in case of lower bainite, the carbides are generally precipitated at 55°–60° angle to the major bainitic axis as presented in Figure 8.7. It is also termed as plate bainite. It has a martensite-like acicular appearance as it forms as individual plates adopting a lenticular shape. Carbides are on finer scale and have shapes of rods or blades.

8.4.2.1 Formation of Lower Bainite

The rate at which the carbon is diffused from the austenite having low carbon content is the rate-controlling step in case of formation of lower bainite. The nature of carbide formed is a function of the chemical composition of the steel and the temperature at which transformation takes place.

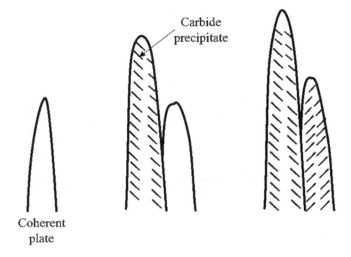

FIGURE 8.7 Different stages involved in the formation of lower bainite.

8.5 MARTENSITIC TRANSFORMATION

Martensite is formed by quenching austenite. Martensitic transformation is a diffusionless transformation. Since there is no diffusion of carbon in this transformation, there is no change in the chemical composition, i.e., both alpha and gamma irons are having the same composition. Hence, α-iron becomes saturated with carbon as it is well known that the solubility limit of carbon is considerably more in γ-iron than that of α-iron. Accordingly, martensite is usually defined as a supersaturated solid solution of carbon in α-iron. Martensite has a body-centered tetragonal structure. The tetragonality is due to the trapping of carbon in the solution. While two dimensions of the unit cell are same, the third one is slightly expanded because of the trapped carbon.

There are several other important characteristics of the martensitic transformation, which are described in the following.

8.5.1 Important Characteristics

8.5.1.1 Diffusionless Transformation

Martensitic transformation is a diffusionless transformation, as martensite has the same composition as the parent austenite. Carbon remains in dissolved solid solution state in martensite, even if the other alloying elements are present.

8.5.1.2 Surface Relief

Martensite crystal is displaced partly above and partly below the surface of the parent austenite. It indicates that the shear (or displacive) mechanism is responsible for this transformation. Surface relief is an essential characteristic of martensitic transformation.

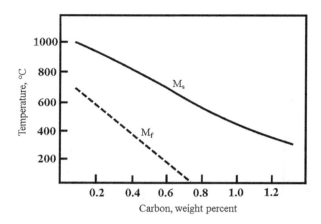

FIGURE 8.8 Effect of carbon in the martensite range.

8.5.1.3 M_s–M_f Temperatures

Martensitic transformation starts at M_s temperature and ends at M_f temperature. The M_s temperature is a specific temperature for a given steel. This temperature depends on the chemical composition of the steel. The relationship between M_s temperature and the chemical composition is as shown in the Andrews equation:

$$M_s(°C) = 539 - 423(\%C) - 30.4(\%M_n) - 12.1(\%Cr) - 17.7(\%Ni) - 7.5(\%M_0) \quad (8.2)$$

Almost all alloying elements, except cobalt and aluminum, lower the M_s temperature. Carbon has the most intense effect on M_s temperature. The higher the carbon content of steel, the lower the M_s temperature (Figure 8.8). The temperature at which martensitic transformation ends, i.e., M_f, is a function of the cooling rate and, hence, can be changed by controlling the cooling rate. The M_f temperature is usually lowered by reducing the rate of cooling. Intuitively, this can be attributed to the fact that M_f is mostly not mentioned in the TTT diagrams.

8.5.1.4 Athermal Transformation

The transformation of austenite to martensite proceeds on continuous cooling below the M_s temperature and stops as soon as cooling is interrupted, i.e., the steel is allowed to stay at a temperature between the start and finish temperatures of martensitic transformation.

Further transformation is possible only upon further cooling. Hence, martensitic transformation is usually ascribed as athermal transformation. If the room temperature lies between M_s and M_f temperatures, quenching the steel to room temperature leads to a large amount of untransformed austenite along with martensite. This untransformed austenite is called retained austenite. If the cooling of steel is arrested between M_s and M_f temperatures and again the cooling is resumed by lowering the temperature, the transformation does not start immediately but starts after austenite is undercooled to a much lower temperature, and in the end, a large amount of austenite remains untransformed even after crossing M_s temperature. This extra untransformed austenite because of the arrest is called stabilized austenite.

8.5.1.5 Effect of Applied Stress on Transformation

If stress from outside is applied on austenite above the M_s temperature, the M_s temperature gets raised, i.e., martensite forms above its M_s temperature. The plastically deformed steel may undergo martensitic transformation by cooling below a critical temperature, known as M_d.

For a given plastic deformation, the percentage of martensite formed becomes higher as the transformation temperature is brought down from M_d to M_s.

8.5.2 Hardness of Martensite

The presence of carbon is the main reason behind the high hardness of martensite in the case of plain carbon steels. It increases as carbon content in the steel increases. Both M_s and M_f temperatures are controlled by the steel composition, which eventually decide the extent of hardening of the martensite. With lowering in carbon content, the M_s and M_f temperatures are increased. Hence, in high-carbon steel, the retained austenite content usually remains at a higher side.

8.5.3 Morphology of Martensite

8.5.3.1 Lath Martensite

It has the shape of the strip, the length of which has the largest dimension and is limited by the grain boundary of austenite. Laths are grouped in packets in parallel fashion laths and have a high density of dislocations in the range of 10^{15}–10^{16} m^{-2} as in a heavily cold worked metal and are created by the process of slip (Figure 8.9a). Lath martensite is formed when M_s temperature is high, such as in low and medium carbon steels.

8.5.3.2 Plate Martensite

Plate martensite is also called acicular or lenticular martensite (lens-shaped). It resembles the shape of mechanical twins, as shown in Figure 8.9b. It is formed in steels having low M_s temperatures, i.e., steels having higher carbon as well as alloying elements.

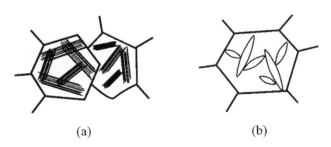

FIGURE 8.9 Schematic micrograph of (a) lath martensite and (b) plate martensite.

FURTHER READING

Singh, V. *Physical Metallurgy.* (Standard Publishers Distributors, New Delhi, 2005).

Avner, S. H. *Introduction to Physical Metallurgy.* (Tata McGraw-Hill Education, New York, 1997).

Sharma, R. C. *Principles of Heat Treatment of Steels.* (New Age International, New Delhi, 2003).

9 Heat Treatment Furnaces

9.1 INTRODUCTION

The achievement of excellent heat treatment properties relies upon the appropriate choice of heat treatment furnaces and the atmosphere maintained inside the furnace. Heat treatment furnaces mostly contain heating chambers in the form of refractory vessels that hold the charge and contain the heat. There are different sources of heat through which the temperature inside the furnace is controlled. Depending upon the size of the charge, heating mechanism, charge handling system, and several other parameters, heat treatment furnaces are designed.

9.2 CLASSIFICATION OF FURNACES

The heat treatment furnaces are generally classified according to the source of heat, use, type of operation, and working environment.

Based on the sources of heat, the furnaces can be classified as fuel-fired furnaces and electrically heated furnaces. The choice of the right fuel depends on the availability and cost of fuel. Depending on the type of fuel, the fuel-fired furnaces can be further classified as solid fuels, liquid fuels, and gaseous fuels. Commonly used solid fuels are coal, pulverized coal, and coke. The advantage of using coal as a solid fuel is its cheap availability, although smoke and lack of temperature control are associated with it as limitations. The problem of smoke can be avoided by using coke since it is less volatile and has a calorific value less than that of coal. Among liquid fuels, fuel oil, gasoline, and kerosene are typically used. These liquid fuels can be easily stored and fired at any time, but they are not being used nowadays due to their high cost. Nevertheless, oil-fired furnaces are used where a high temperature (>1000°C) is required. A heat circulation arrangement is necessary for such types of furnaces, as the temperature inside the furnace is not uniform, and hence, these are not economical. On the other hand, gas-fired furnaces are economical and have specific advantages over other furnaces. They possess better control of temperature, have a simpler design, and can be used up to 1500°C. Some examples of gaseous fuels are natural gas, coal gas, producer gas, cracked oil gas, and refinery gas.

Electrically heated furnaces offer certain advantages over other furnaces such as more straightforward design, uniformity of temperature inside the furnace chamber, highest efficiency in heat utilization, and clean working conditions. These are the furnaces that can attain very high temperatures and are convenient to start and switch off at any moment. Based on the method of heating, electric furnaces are again classified as resistance furnace, induction furnace, plasma arc furnace, arc furnace, and electron beam furnace. Resistance furnaces are typically used for the heat treatment of metals and alloys. The temperature in such types of furnaces can

TABLE 9.1
Common Resistors Used in Electrical Resistance Furnaces

Name	Composition	Maximum Working Temperature (°C)
Constantan	Cu, 40% Ni	900
Nichrome I	Ni, 20% Cr	1100
Nichrome II	Ni, 24% Fe; 16% Cr	950
Alumel	Ni, 3% Mn; 2% Al; 1% Si	1200
Chromel	Ni, 10% Cr	1200
Chromel C	Ni, 23% Fe; 15% Cr; 2% Mn	900
Kanthal	Fe, 25% Cr; 5% Al; 3% Co	1400
Tungsten		2400
Molybdenum		1800
Tantalum		2200
Platinum		1500
Pt–Rh alloy, Pt, 10% Rh		1700
Thoria		2400
Graphite		2000

be controlled easily to a high degree of accuracy. Various resistors are used in these furnaces depending upon the desired temperature. A list of resistors used in electrical resistance furnaces is given in Table 9.1. Induction furnaces are generally used for surface hardening purposes. Other furnaces such as plasma arc furnace, arc furnace, and electron beam furnace are mainly used for melting metals and alloys.

Depending on the use or heat treatment types, the furnaces can also be classified. For example, salt bath furnaces are used for treatments such as annealing, normalizing, and hardening. Sealed quench furnaces are used for carburizing or carbonitriding purposes. According to the type of operation, the furnaces are categorized into two groups, namely, batch furnace and continuous furnace. Both these types of furnaces are described in detail in the next two sections.

9.3 BATCH FURNACE

A batch furnace consists of a refractory-lined insulated chamber enclosed in a steel shell with one or more access doors. These furnaces are widely accepted because of their flexibility to varying sizes of the workpiece to be heat-treated and different heat treatments to be performed. The term "batch" signifies that the heat treatment of the workpiece is carried out and completed in different batches. A batch furnace can be horizontal or vertical type depending upon the size of the workpiece and type of heat treatment adopted. These types of furnaces are beneficial for carrying out laboratory-scale experiments. Some very commonly used batch furnaces are box-type furnace, muffle furnace, bell furnace, vacuum furnace, fluidized bed furnace, pit furnace, and bogie hearth furnace.

Heat Treatment Furnaces

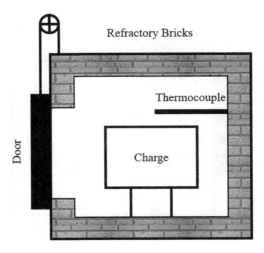

FIGURE 9.1 Schematic diagram of a box-type furnace.

9.3.1 Box-Type Batch Furnace

One of the simplest types of batch furnaces is a box-type batch furnace. The chamber is rectangular in section, and it has an opening just as in a box, as schematically illustrated in Figure 9.1. These furnaces can be used for annealing, pack carburizing, and hardening of low-alloy steels. Generally, the loading and unloading process of the samples is done manually from the front end of the furnace. For heavy components, the bottom loading furnaces are used, in which the bottom cover and the transfer cab are lowered, and the charge is placed on it. The bottom cover is closed after loading of the charge. These box-type furnaces are heated through electricity or fuel.

9.3.2 Muffle Furnace

A muffle furnace is an empty cuboid or tube-shaped furnace made of unique refractory material or nonscaling steel. Any fuel or electrical energy can be used to heat the muffle furnace. In this case, the heat source does not legitimately reach the material being heat-treated. The samples are charged, and electrical energy or gas firing can be utilized to heat the furnace externally. Electrically insulated furnaces are used extensively for the heat treatment of small parts. In electric muffle furnaces, as shown in Figure 9.2, the muffle is surrounded by different heating elements such as SiC heating rods, $MoSi_2$ heating rods, nichrome, or kanthal wires, depending upon the maximum temperature desired.

In contrast, in the case of gas-fired muffle furnaces, the muffle gets heated by the hot gases generated outside the muffle (Figure 9.3). The hot gases are made to circulate through the space between the interior wall and the exterior muffle wall. Since the combustion products of the gas do not enter the heating chamber or muffle, scaling of the components is prevented. Muffle furnaces are used for annealing, nitriding, and hardening purposes.

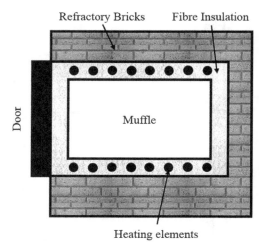

FIGURE 9.2 Schematic diagram of an electric muffle furnace.

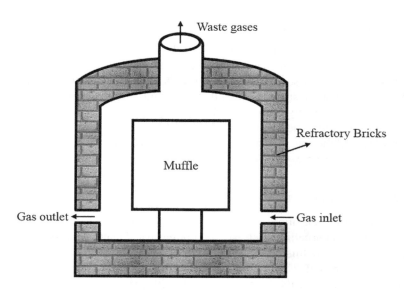

FIGURE 9.3 Schematic diagram of a gas-fired muffle furnace.

9.3.3 Pit Furnace

Pit furnace is mainly used for heavy and large parts such as tubes, spindles, shafts, and rods. It comprises the furnace put in a pit and a cover or lid put over the furnace (Figure 9.4). Most of the portion of the pit furnace lies below the ground level. Fans placed inside the pit furnace promote uniformity of temperature and gas composition, thereby reducing distortion and warpage. Pit furnaces are especially appropriate

Heat Treatment Furnaces

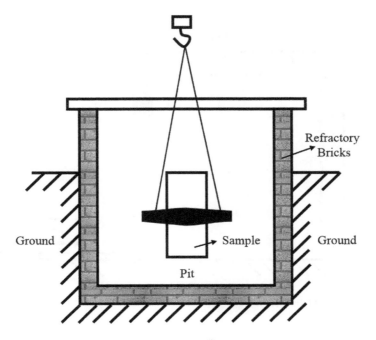

FIGURE 9.4 Schematic diagram of a pit furnace.

for parts to be cooled in the furnaces as direct quenching with such huge charge, and the enormous furnace is not possible.

9.3.4 Bogie Hearth Furnace

It is a modified variant of the box-type batch furnace. The furnace is especially suitable for the heat treatment of large and heavy components. The bogie acting as a hearth with the full charge is put inside the furnace, which is refractory topped. A schematic illustration of the bogie hearth furnace is shown in Figure 9.5. The heating is done by a fuel or by some electric resistance heating elements. Proper sealing controls the atmosphere inside the furnace. These types of furnaces are typically used for stress-relieving, annealing, and hardening purposes.

9.3.5 Bell Furnace

Bell furnaces, also referred to as liftoff cover furnaces, consist of two containers (Figure 9.6). The inner container comprises the charge, which is placed on the hearth, and a retort on the top of the charge. The retort is sealed at the base with sand. The outer bell-shaped container, which carries the heating elements, covers the inner container. Protective gas is supplied to the sealed retort for constant protection. After heating of the first charge, the bell-shaped container is lifted off and placed on the second assembly to carry out the heating process. In the meantime,

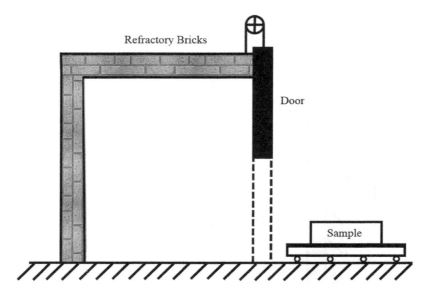

FIGURE 9.5 Schematic diagram of a bogie hearth furnace.

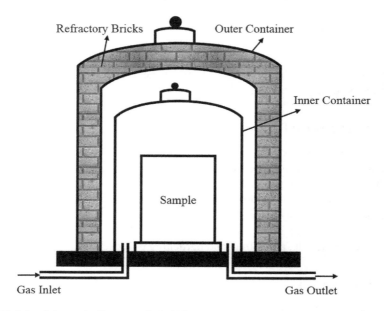

FIGURE 9.6 Schematic diagram of a bell furnace.

the first assembly and its charge cool down under the protective atmosphere, as the gas is continuously being fed all through the sealed retort. Such types of furnaces are widely used in wire industries for annealing coils and strips under a controlled atmosphere to avoid decarburization or oxidation.

9.4 CONTINUOUS FURNACE

In a continuous furnace, the parts to be heat-treated are continuously introduced from one end of the furnace. After the heat treatment of the parts is over, the heat-treated parts are forced to move to the other end of the furnace where they are discharged. These furnaces are meant for typically mass production of components and are, therefore, mostly found in industries or pilot plants of national laboratories. A continuous furnace can have different zones depending upon the type of heat treatment adopted. For example, a continuous carburizing furnace has separate chambers for heating, carburizing, and the diffusion process. A rotary kiln has different zones for preheating, reaction, and cooling purposes. Modern continuous furnaces include rotary hearth furnace, conveyor furnace, tunnel furnace, shaker hearth furnace, rotary retort furnace, and rotary kiln.

9.4.1 ROTARY HEARTH FURNACE

The rotating hearth furnace consists of a rotating hearth that rotates along its vertical axis inside a stationary roof, as shown in Figure 9.7. Components with different compositions can be heat-treated in such types of furnaces. The components to be heat-treated are charged through an opening and discharged from the same opening or adjacent opening after the completion of the heat treatment cycle. The speed of rotation is varied to such an extent that the heat treatment cycle is finished by the time the hearth finishes one complete revolution.

9.4.2 CONVEYOR FURNACE

In conveyor furnaces, endless belts are utilized to move the segments through the heater. Such furnaces are suitable for smaller parts. The parts are charged continuously into the belt at one end of the furnace and discharged from the other end. Inside the furnace, the movement of the belts carrying the parts is regulated in such a way

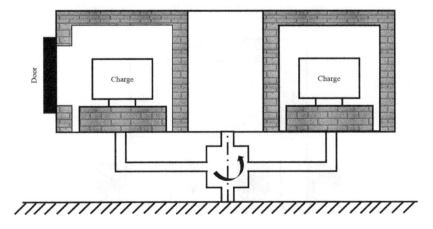

FIGURE 9.7 Schematic diagram of a rotary hearth furnace.

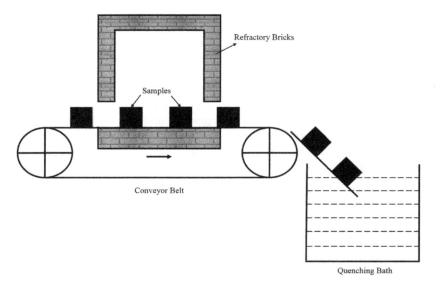

FIGURE 9.8 Schematic diagram of a conveyor furnace.

that the heat treatment is completed by the time the parts are discharged. Figure 9.8 shows the schematic illustration of the conveyor furnace. These furnaces are applicable for solidifying, hardening, tempering, and normalizing purposes.

9.4.3 Tunnel Furnace

In a tunnel furnace, as shown in Figure 9.9, several cars carrying the charge are moved into the furnace at one end. The cars are pushed through the furnace with the help of some mechanical means. The movement of cars is so arranged that when a car is removed at the discharged end, another car is loaded simultaneously at the charging end. These types of furnaces are typically used for annealing purposes.

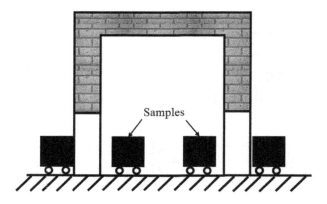

FIGURE 9.9 Schematic diagram of a tunnel furnace.

Heat Treatment Furnaces 125

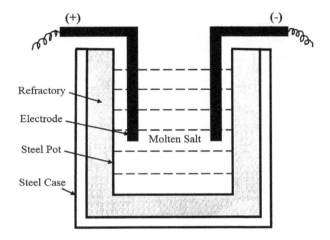

FIGURE 9.10 Schematic diagram of a salt bath furnace.

9.5 SALT BATH FURNACE

A salt bath furnace (Figure 9.10) consists of an oval or rectangular container made of ceramic or metal. This container holds the molten salt in which the samples to be heat-treated are immersed. The container is heated and maintained at a required temperature either by the combustion of fuel or by electric resistors. Nitrates, chlorides, carbonates, cyanides, and caustic soda are the most commonly used salts in this furnace. The heat transfer takes place very quickly, as salt bath comes in the best intimate contact with the charge since the mode of heat transfer is mainly by convection. Also, the molten salts possess high heat capacity, resulting further in rapid heating of the charge as compared to conventional furnaces. Thus, heat treatment time is reduced, resulting in a good economy. Salt bath furnaces can be used for various heat treatment operations like austempering, martempering, hardening, cyaniding, carburizing, and tempering. Salt bath furnaces hold certain advantages over other furnaces. Oxidation and decarburization can be avoided easily, as the sample is in direct contact with the molten bath. Worn out electrode can be easily replaced while the furnace is in operation. Complicated shapes with variable section thickness can also be heated along with the simply shaped object. In the same way, light and heavy objects can be handled in the same bath.

FURTHER READINGS

Singh, V. *Heat Treatment of Metals*. (Standard Publishers Distributors, New Delhi, 2006).
Rajan, T. V., Sharma, C. P., & Sharma, A. *Heat Treatment: Principles and Techniques*. (PHI Learning Pvt. Ltd., New Delhi, 2011).

10 Heat Treatment Atmosphere

10.1 INTRODUCTION

Heat treatment furnace atmospheres have played an essential part in successful heat treatment practices. The furnace atmosphere that protects the components from oxidation or decarburization during the process of heat treatment and also maintains the required surface properties of the components is called a controlled atmosphere. The composition and distribution of the furnace atmosphere are adjusted or controlled to achieve specific surface properties. Depending on the type of heat treatment adopted, controlled atmospheres can be either protective or chemically active.

A protective atmosphere protects the material from chemically reacting with harmful elements, which may lead to oxidation or decarburization during heat treatment. Examples of heat treatment practices where protective atmospheres are used include bright annealing, normalizing of ferrous and nonferrous metals, and annealing of malleable cast iron. In contrast, in the case of chemically active atmospheres, the surface chemically reacts with the atmosphere to achieve improved surface characteristics. Examples include carburizing and carbonitriding of steel components, decarburization, nitriding, and hardening processes.

Several gases are used in heat treatment furnaces as controlled atmospheres, including hydrogen, nitrogen, helium, argon, ammonia, propane, carbon monoxide, carbon dioxide, methane, and butane. In some cases, a mixture of gases is used, which decides the final properties of the controlled atmosphere. The inert gases, particularly argon and nitrogen, are used as protective gases that prevent unwanted chemical reactions like oxidation or decarburization from occurring on the metal surface.

10.2 REACTIONS BETWEEN ATMOSPHERE AND MATERIAL

The reactions that occur between the controlled atmosphere containing gases and the material are classified into three categories: the reaction between oxygen and the material, the reaction between carbon and the material, and the reaction between gases.

10.2.1 Reactions between Oxygen and Material

The reactions between oxygen and the material to be heat-treated form the basis for protecting the materials from oxidation. Hence, understanding of these reactions is very much essential. The reactions that control oxidation in furnaces, like in the case of bright annealing of low-carbon steel, are as follows:

$$\text{Material} + H_2O \rightleftharpoons \text{Oxide} + H_2 \tag{10.1}$$

$$\text{Material} + CO_2 \rightleftharpoons \text{Oxide} + CO \tag{10.2}$$

The equilibrium constants for the above reactions can now be written as

$$K_1 = \frac{p_{H_2}}{p_{H_2O}} \tag{10.3}$$

$$K_2 = \frac{p_{CO}}{p_{CO_2}} \tag{10.4}$$

Hence, the equilibrium constant at temperature T, K_T, is given as

$$K_T = \frac{K_1}{K_2} = \frac{p_{CO} \times p_{H_2O}}{p_{CO_2} \times p_{H_2}} \tag{10.5}$$

The equilibrium constant K_T is also related to temperature as follows:

$$\log_{10} K_T = -\frac{\Delta G^0}{4.576T} \tag{10.6}$$

where ΔG^0 is the free energy of the reaction. Therefore, it can be seen that the oxidation/reduction potential depends not only on K_1 and K_2 but also on the oxide dissociation pressure of the particular material at the given temperature.

10.2.2 Reactions between Carbon and Material

Carburization and decarburization are the two basic heat treatment methods where reactions between the carbon and the material are involved. While carbon monoxide and methane are the gases used for carburizing, the gases responsible for decarburization are carbon dioxide, water vapor, and hydrogen. The reactions involved are as follows:

$$2CO + 3Fe \rightleftharpoons CO_2 + Fe_3C \tag{10.7}$$

$$CH_4 + 3Fe \rightleftharpoons 2H_2 + Fe_3C \tag{10.8}$$

Heat Treatment Atmosphere 129

The equilibrium constants for the above reactions can now be written as

$$K_3 = \frac{p_{CO_2}}{p^2_{CO}} \tag{10.9}$$

$$K_4 = \frac{p^2_{H_2}}{p_{CH_4}} \tag{10.10}$$

Therefore, at any given temperature, the $\dfrac{p^2_{CO}}{p_{CO_2}}$ or $\dfrac{p_{CH_4}}{p^2_{H_2}}$ ratios can be used to assess the carbon potential of the atmosphere. In the case of decarburization, water vapor plays a vital role, which is evident from the following reaction:

$$H_2O + Fe_3C \rightleftharpoons H_2 + CO + 3Fe \tag{10.11}$$

Another reaction (equation 10.12) is very significant for nonferrous metals where the atmosphere is contaminated with sulfur.

$$H_2S + Metal \rightarrow H_2 + Metal\ Sulphide \tag{10.12}$$

10.2.3 Reactions between Gases

Gases like oxygen, carbon, and hydrogen are often used to obtain equilibrium compositions at various temperatures. For example, the equilibrium is established by the following reaction at temperatures above 800°C:

$$CO + H_2O \rightleftharpoons CO_2 + H_2 \tag{10.13}$$

It may be noted that carbon monoxide has a higher affinity for oxygen at lower temperatures, whereas hydrogen has a higher affinity for oxygen at higher temperatures. The affinities of both these gases toward oxygen are equal at about 850°C. Because of the water–gas reaction (equation 10.13), the concentration of carbon monoxide is reduced at low temperature wherever the atmosphere contains water vapor.

10.3 TYPES OF FURNACE ATMOSPHERES

10.3.1 Exothermic Atmosphere

Exothermic atmospheres are produced by the exothermic combustion of gases and air and are low-cost prepared furnace atmosphere. Depending upon the air-to-gas ratio, exothermic atmospheres are again subgrouped to lean (totally burnt) and rich (flammable) types. Nominal composition in volume percentage for the rich atmosphere is

$$N_2 = 71.5\%;\ CO = 10.5\%;\ H_2 = 12.5\%;\ CH_4 = 0.5\%$$

130 Phase Transformations and Heat Treatments of Steels

The lean atmosphere generally contains 0%–3% CO, 0%–4% H_2, and rest N_2. The nominal composition in volume percentage is

$$N_2 = 86.8\%; CO = 1.5\%; H_2 = 1.2\%; CO_2 = 10.5\%$$

It principally constitutes nitrogen and, therefore, is used as an inert atmosphere in most of the heat treatment processes.

Rich atmospheres are commonly used for bright annealing, normalizing, and tempering of steel. On the other hand, lean atmospheres are employed for the heat treatment of nonferrous metals and alloys like bright annealing of copper. Exothermic atmospheres cannot be used where decarburization is involved. It is because these atmospheres have low carbon potential due to the presence of carbon dioxide. Therefore, gases like water vapor and carbon dioxide are removed to produce exothermic atmospheres. Examples include pure nitrogen and nitrogen-based atmospheres containing carbon monoxide and hydrogen up to 25% of the total.

10.3.2 ENDOTHERMIC ATMOSPHERE

Endothermic atmospheres are produced when a hydrocarbon-containing fuel reacts with oxygen to oxidize the hydrocarbon to carbon monoxide and hydrogen. The reactions involved are as follows:

$$2CH_4 + O_2 + 4N_2 = 2CO + 4H_2 + 4N_2 \qquad (10.14)$$

$$2C_3H_8 + 3O_2 + 12N_2 = 6CO + 8H_2 + 12N_2 \qquad (10.15)$$

The endothermic gas atmosphere can be used in any furnace where reducing conditions are required, but it is typically used as a carrier gas in gas carburizing and carbonitriding applications. Endothermic generators find applications in heat treatment and brazing of carbon steels, sintering, and as a base gas for gas carburizing and carbonitriding of steels.

10.3.3 AMMONIA-BASED ATMOSPHERE

The anhydrous liquid ammonia is dissociated into its constituents by heating in an electrically heated or gas-fired chamber having one or more catalyst (iron at 560°C or nickel at 900°C). The following reaction takes place during the dissociation:

$$2NH_3 \rightarrow N_2 + 3H_2 \qquad (10.16)$$

The product of the above reaction is called cracked ammonia that comprises 75% hydrogen and 25% nitrogen. Based on the grade of ammonia used, the dew point of the cracked ammonia gas varies from −15°C to −40°C.

Burnt ammonia is produced by passing ammonia or air mixture over a catalyst at 850°C. The reaction that occurs during the burning is as follows:

$$4NH_3 + 3O_2 + 12N_2 \rightarrow 14N_2 + 6H_2O \qquad (10.17)$$

Heat Treatment Atmosphere

Burnt ammonia is generally used for bright annealing of strip and wire, whereas cracked ammonia is used for bright annealing of stainless steels, sintering, and other applications.

10.3.4 CHARCOAL-BASED ATMOSPHERE

Such type of atmosphere is prepared by burning charcoal with air or flue gas. Air is passed through a bed of hot charcoal, which burns the charcoal to N_2, CO_2, and H_2O at the bottom-most part of the bed. The highly heated charcoal then converts CO_2 to CO and H_2O to H_2 gas as the gas mixture moves up the bed. These atmospheres find applications in annealing, normalizing, and hardening high-carbon steels without scale or decarburization. Because of high operating costs and the inability to make it automatic, these atmospheres are less used.

10.3.5 HYDROGEN ATMOSPHERE

Hydrogen is commercially available at the purity of 98%–99.9% with traces of water vapor and oxygen. It is obtained by the electrolysis of water, or by the decomposition of ammonia, or by the catalytic conversion of hydrocarbons, or by the water–gas reaction. Hydrogen produced by the electrolysis of distilled water is best suited for metallurgical purposes. Hydrogen is a powerful deoxidizer, but in the dry state, it decarburizes high-carbon steels to form methane at high temperatures. Hydrogen, when adsorbed, causes hydrogen embrittlement, especially in high-carbon steel. Dry hydrogen is often used as an atmosphere in annealing of low-carbon steels, stainless steels, direct reduction of metal ores, and sintering of metal powder components.

FURTHER READING

Singh, V. *Heat Treatment of Metals.* (Standard Publishers Distributors, New Delhi, 2006).
Rajan, T. V., Sharma, C. P., & Sharma, A. *Heat Treatment: Principles and Techniques.* (PHI Learning Pvt. Ltd., New Delhi, 2011).

11 Common Heat Treatment Practices

11.1 INTRODUCTION

A metal or alloy must be tuned to the desired properties before use in several engineering applications. Most of these properties depend upon the structure, and suitable heat treatment processes can produce the preferred phases. According to *Metals Handbook*, heat treatment may be defined as "a judicious combination of both heating and cooling operations, including heating/cooling rate, heat treatment temperature performed on metals or alloys in their solid-states in a synchronized way that will yield desired properties." The initial step is heating the material to some temperature to form austenitic structure and then holding at that temperature for the required time, followed by cooling down to room temperature. These transformed products define the physical and mechanical properties of the material. Therefore, looking at the importance of a detailed understanding of the heat treatment process, this chapter mainly focuses on different heat treatment practices that are commonly used for steels since steel is one of the most essential and versatile materials for numerous engineering applications. Subsequent sections describe various heat treatment methods for other metals and alloys.

11.2 TYPICAL HEAT TREATMENT PROCESSES

11.2.1 ANNEALING

In simple words, annealing treatment comprises subjecting the steel to a predefined austenitic temperature and holding for some time depending on its thickness followed by slow cooling, preferably furnace cooling. The treatment method depends not only on the steel composition but also on other dimensional parameters such as size, shape, and ultimately the final properties desired. The annealing treatment serves numerous purposes; some of them are as follows:

- Relieve internal stresses generated due to casting, welding, or other mechanical working operations
- Better machinability
- Achieve chemical uniformity
- Restore toughness of the metallic component
- Refinement of grain size
- Decrease the gaseous contents of steel

133

11.2.1.1 Full annealing

The term annealing, without any prefix or suffix, usually means full annealing. It comprises heating the specimen to a high temperature, normally above its upper critical temperature, holding for a sufficient time so as to ensure homogenous austenite, and finally, slow cooling that is usually performed in the furnace. Steel specimen for this purpose is heated to a temperature, which is usually around 20°C–40°C higher than the upper critical temperature (A_3) in case of hypoeutectoid steels. Heating at this temperature results in austenite with finer grains, which upon slow cooling below A_1 yields ferrite and pearlite of fine grains structure. On the other hand, if the austenitizing temperature is about 100°C above A_3, grain growth of austenite occurs, and the coarse grains of austenite are not refined upon cooling, which in turn impairs the properties of the steel. The previous discussions are illustrated in Figure 11.1.

Steel is heated to around 20°C–40°C above the lower critical temperature (A_1) in case of a hypereutectoid steels. Heating at this temperature results in fine grains of austenite with partly or entirely spheroidized Fe_3C, which on slow cooling through A_1 produces fine grains of pearlite and spheroidized cementite. The reason behind such spheroidization of cementite may be ascribed to the reduction in the interfacial area between austenite and cementite, giving rise to a lower interfacial energy. Heating slightly above A_{cm} temperature results in a single-phase homogenous austenite.

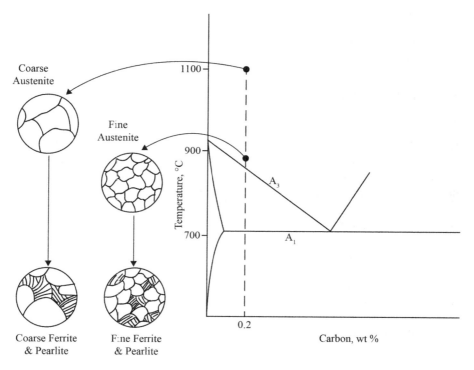

FIGURE 11.1 Schematic illustration of microstructural changes on annealing of a typical hypoeutectoid steel (0.2% carbon steel).

Common Heat Treatment Practices

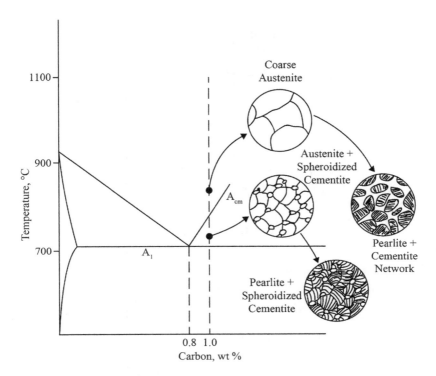

FIGURE 11.2 Schematic illustration of microstructural changes on annealing of hypereutectoid steel (1% carbon steel).

Such steel on slow cooling gets proeutectoid cementite formed at the grain boundaries of austenite as a thicker network, and pearlite has coarse grains, as shown in Figure 11.2. Generally, heating to such temperature ranges for annealing is avoided because of the following reasons:

- The cementite (which is very brittle) network facilitates brittle fracture by providing path for crack propagation
- Grain coarsening of austenite occurs
- High energy consumption
- More time and less productivity
- More scaling and decarburization with no advantage

11.2.1.1.1 Purpose of Full Annealing
- Steel castings, in particular, have invariably coarse austenite grains, which result in coarse ferrite or pearlite grains, commonly called as "Widmanstatten" structures (Figure 11.3). These structures have feeble impact strength and low toughness. Hence, annealing is performed in such steel castings or hot worked steels to improve the mechanical properties by refining the grain size.

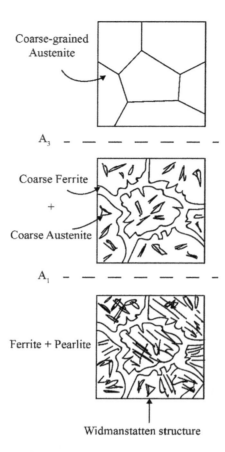

FIGURE 11.3 Formation of Widmanstatten structure in steel castings.

- Annealing is also done to soften the steel. The slow cooling causes the transformation of austenite at A_1 or close to it to coarse lamellar pearlite, which is very soft and, therefore, results in softening of steel with a reduction in hardness.
- Internal stresses of different types, also known as residual stresses, are usually developed because of high strains associated with welding, cold working, grinding, machining, case hardening, phase transformation, and so on. Annealing is done to relieve these internal stresses. Slow cooling during and after phase transformation induces no internal stresses.
- Steels with 0.3%–0.4% carbon have better machinability in the annealed state. As annealing results in coarse pearlite and ferrite, which increase the softness and ductility, it results in improved machinability.
- Annealing reduces some defects such as aligned sulfide inclusions or bands in steels.

Common Heat Treatment Practices

11.2.1.2 Homogenizing Annealing/Diffusion Annealing

Typically, heavy plain carbon steel castings, steel ingots, and high-alloy steel castings tend to have columnar grains, dendrites, and nonhomogeneous structures, which render brittleness or, in other words, decrease the ductility and toughness of steel. Homogenizing annealing is used to get rid of such nonuniform structures in steels. In this process, the steel component is heated much above the upper critical temperature and is soaked for prolonged periods at that temperature, followed by slow cooling. As the diffusion of most of the alloying elements in steels is much lower than carbon, the alloy steel ingots are usually homogenized at higher temperatures of 1150°C–1200°C for 10–20h, followed by slow cooling. The hypoeutectoid inhomogeneous alloy steels are held at 1000°C for 1–2h, whereas hypereutectoid alloy steels are held for 5–6h. Homogenization impairs the properties because it coarsens the austenite grains. Thus, steels after such heat treatment undergo either normalizing or full annealing.

Homogenization annealing also produces thick scales on the steel surfaces. Scaling results in loss of metal. A controlled heating thus is a recommended method where a precise temperature is fixed and the required soaking period is considerably low. Higher temperatures, prolonged holding time, slow cooling rates in addition to higher amount of scaling, and further heat treatment make this process highly economically expensive.

11.2.1.3 Partial Annealing

It is also termed as intercritical annealing or incomplete annealing. It is comprised of subjecting the steel to an intermediate temperature between the lower critical temperature (A_1) and upper critical temperature $(A_3$ or $A_{cm})$. Subsequent cooling has to be carried out slowly till room temperature or may be in air from an intermediate temperature such that all the austenite gets converted to pearlite.

11.2.1.4 Subcritical Annealing

It is a heat treatment operation in which the maximum heating temperature is always below the lower critical temperature (A_1). Here, no phase transformation is expected to take place. This is only a thermally activated phenomenon, which facilitates processes like recovery, recrystallization, grain growth, and agglomeration of carbides.

11.2.1.5 Recrystallization Annealing

Practically, all steels that are heavily cold-worked are processed by this annealing. In this process, the steel component is heated just above its recrystallization temperature followed by holding for homogenization and subsequent cooling. It results in improved ductility at the expense of hardness or strength. The final microstructure now consists of fine equiaxed ferritic grains that are almost strain-free. The primary target for performing a recrystallization annealing is to restore ductility and toughness and refinement of coarse grains and to improve the electric and magnetic properties in case of grain-oriented steels. To initiate recrystallization, a critical amount of deformation causing formation of crystalline imperfections is the driving force. These imperfections facilitate initiation of nucleation for formation of new grains. As the amount of imperfections increase, the temperature required for recrystallization tends to decrease. Recrystallization temperature is the temperature at which a cold-worked metal gets 50% recrystallized in 1 h. On average, it is given by the following equation:

$$t_r = (0.3 - 0.5)T_{mp} \tag{11.1}$$

where t_r is the recrystallization temperature and T_{mp} is the melting temperature, both in the Kelvin scale. There are some essential points to remember related to recrystallization temperature, which is summarized in the following:

- Recrystallization temperature increases with the purity of the metal. Pure metals always have low recrystallization temperature than impure metals.
- Zinc, tin, and lead possess recrystallization temperatures below the room temperature. Hence, these metals cannot be cold-worked at room temperature since they recrystallize spontaneously, reforming a strain-free lattice structure.
- Recrystallization temperature decreases with increasing annealing time.
- The finer the initial grain size, the lower the recrystallization temperature.

11.2.1.5.1 *Hot and Cold Working of Metals*

When a material is worked or deformed plastically above its recrystallization temperature, the process is called as hot working. On the other hand, cold working is the process of plastic deformation of material below the recrystallization. Let us understand these two processes in line with the process of plastic deformation. Two opposing phenomena cooccur when a material is plastically deformed at a high temperature. One is the hardening phenomenon, which happens due to the plastic deformation, and the other is the softening phenomenon, which happens due to the recrystallization. At a specific temperature, these two cooccurring phenomena balance themselves, and the process of working the material above this specific temperature is known as hot working, whereas cold working is nothing but working below this temperature. Examples of hot working include working of lead and tin at room temperature since their recrystallization temperature is below the room temperature. However, working a metal like tungsten at 1200°C is still considered as cold working, as its recrystallization temperature is higher than 1200°C.

11.2.1.6 Stress-Relieving Annealing

Internal stresses (residual stresses or locked-in stresses) are generated in the course of different operations like welding, solidification, grinding, machining, shot peering, case hardening, and precipitation. Tensile stresses, particularly in surface layers, are most dangerous to cause warpage or even cracks at low or without any external stresses. When steel is heated to below A_1 temperature to eliminate such type of residual stresses, it is then called stress-relieving annealing. The primary aim is to remove the harmful tensile stresses, allow higher external loads, increase fatigue life and prevent intercrystalline corrosion, increase the impact resistance and lower the

Common Heat Treatment Practices

susceptibility to brittle fracture, reduce the chance of cracking, prevent stress corrosion, and achieve dimensional stability.

11.2.1.7 Isothermal Annealing

Steel austenitized at temperatures of 20°C–40°C above A_3 is cooled quickly to the temperature of isothermal holding, which is below A_1 temperature in the pearlitic range. As a result, all the austenite is converted into pearlite. Although the target of this isothermal annealing is more or less similar to that of full annealing, it possesses certain advantages over full annealing. The process of isothermal annealing is cheaper than full annealing because of a shorter heat treatment cycle. As at a constant temperature the transformation takes place, the microstructure obtained in case of isothermal annealing is more uniform as compared to full annealing where transformation occurs over a range of temperatures. The hardness of steel can be thus controlled better in isothermal annealing. Isothermal annealing improves machinability with an excellent surface finish.

11.2.1.8 Patenting

Patenting is an application of isothermal transformation, which consists of austenitizing steel in a continuous furnace to temperatures of 150°C–200°C above AC_3 (usually 870°C–930°C) to get completely homogenous austenite. The steel is then soaked for a fairly long period of time and then quenched in a bath (lead or salt bath) maintained at a constant temperature. The typical temperature of the bath varies from about 450°C to 550°C. For steels, the temperature of quenching medium is chosen close to the nose of the time–temperature transformation (TTT) curve, which results in the transformation of austenite to pearlite. Once the transformation is complete, the steel is cooled down to room temperature by either keeping in air or water spraying.

In some cases, a minor amount of upper bainite can be observed. Eutectoid steel after patenting may have an interlamellar spacing of 40 nm of strength 1240–1450 MPa, which is cold drawn to the interlamellar spacing of 10 nm. Fine interlamellar spacing easily blocks the motion of dislocations, increasing the strength drastically with high toughness in twisting and bending. In most practical situations, this patenting process is used for producing springs, high strength ropes, and piano wires of usually 0.45%–1.0% carbon steel.

11.2.1.9 Spheroidization Annealing

It is the method of heating to a specific temperature, typically the austenitizing temperature in most cases, and holding at that temperature followed by prolonged cooling to produce spheroidal pearlite or carbides in globular forms in steels. Usually, the austenitizing temperatures used for different types of steels are as follows:

Eutectoid steels – 750°C–760°C
Hypoeutectoid steels – 770°C–790°C
Hypereutectoid steels – 770°C–820°C
High-speed steels – 875°C

The main aim of obtaining spheroids structure is to obtain maximum softness, ductility, and machinability with minimum hardness in the material. It is applied to high-alloy tool steels and high-carbon steels to improve machinability and ductility. Low-carbon steels may be spheroidized for cold forming, such as tubing. A microstructure of coarse spheroidized cementite (or alloy carbides) particles embedded in a ferrite matrix is generally seen in this case.

Various types of annealing treatments are discussed, and each treatment has a different heating profile depending upon the application areas. The temperature ranges for these different annealing processes are illustrated in Figure 11.4 for better understanding.

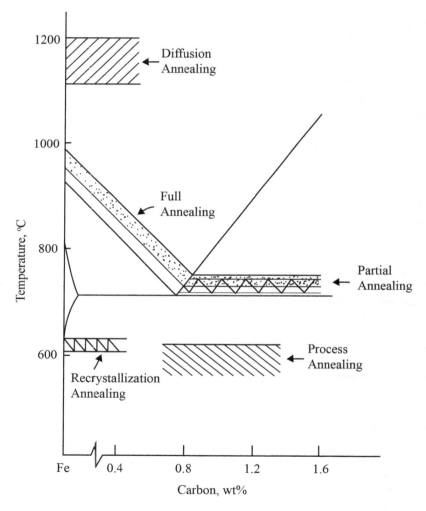

FIGURE 11.4 Different types of annealing treatments and their respective temperature ranges.

Common Heat Treatment Practices

11.2.2 Normalizing

It is another heat treatment process, where the steel is taken to a temperature above the upper critical temperature (A_3 or A_{cm}), soaked there for a proper time followed by air cooling. The austenitizing temperature ranges for various steels are

$$T_{normalizing} = A_3 + \left(40°C \text{ to } 60°C\right) \text{ for hypoeutectoid steels and eutectoid steel}$$

$$= A_{cm} + \left(30°C \text{ to } 50°C\right) \text{ for hypereutectoid steels}$$

It can be seen that the temperature for the normalizing treatment is higher than that for annealing. It increases the homogeneity of austenite, thereby leading to a better dispersion of ferrite and cementite resulting in enhanced mechanical properties. Practically, this treatment is used to refine the coarse grains of steel castings and forgings, which have been not worked at high temperatures. In particular, it eliminates dendritic and nonuniform structures. Normalizing is also used to eliminate or reduce microstructural irregularities. As normalizing reduces the grain size of phases, the resultant microstructure has greater uniformity with excellent mechanical properties. It reduces even bonding in carbon steels. In the case of hypereutectoid steels, it helps break the coarse cementite network. The cementite network does not appear in the normalized structure because of the less availability of time during cooling. A schematic illustration of changes in the microstructures on normalizing at different austenitizing temperatures is shown in Figure 11.5. There are specific other necessary points to remember while conducting the normalizing heat treatment process, which is summarized in the following:

- Austenite formed during normalizing treatment should be homogenous to have excellent stability so that it could be supercooled to temperatures much below A_1 by air cooling before it transforms into fine pearlite.
- Lower is the temperature of transformation; finer is the grain size of ferrite and pearlite with smaller interlamellar spacing.

11.2.3 Hardening

As the title suggests, this treatment is used often to harden the material. In this treatment, the steel component is subjected to an appropriate austenitizing temperature, soaked there to develop homogenous and fine-grained austenite, and then followed by cooling at a faster rate than its critical cooling rates to suppress pearlite and bainitic transformations. Hence, austenite transforms into martensite, and the steel becomes hard; such cooling is called quenching. Usually, carbon steels are quenched in water, and alloy steels in oil. The hardness of the martensite depends mainly on the carbon content of the steel. The critical cooling rate is defined as the slowest cooling rate, which produces a full martensite structure, i.e., the cooling rate touching the nose of the "S" curve. In real cases, this treatment is often used to develop high hardness in tool steels so that it can cut other metals. The cutting ability of tool steel is proportional to its hardness. Besides, many machine parts such as gears,

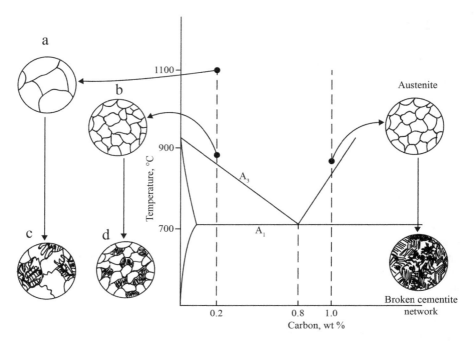

FIGURE 11.5 Schematic illustration of changes in the microstructure on normalizing. (a) Higher austenitizing temperature results in coarse austenite grains (more homogeneous). (b) Austenitizing in normalizing range results in slight coarser grains than that obtained in annealing range. (c) Normalizing results in Widmanstatten structure. (d) Finer ferrite and pearlite is obtained by normalizing at a lower temperature.

shafts, cams, and bearings are hardened to induce high wear resistance. Another essential objective of hardening is to develop high yield strength in machine components made of structural steels with excellent toughness and ductility to bear higher working stresses. The more the yield strength, the higher the magnitude of stress, which a part can carry during service.

11.2.3.1 Hardening Temperature for Different Types of Steels

The hardening temperature depends on the chemical constituents of different kinds of steel and is depicted in Figure 11.6. For example, hypoeutectoid steels are heated to about 20°C–40°C more than the critical temperature (A_3), whereas eutectoid and hypereutectoid steels are subjected to about 20°C–40°C more than the lower critical temperature (A_1). Hypoeutectoid and eutectoid steels, when subjected to the temperature range mentioned earlier, convert to homogenous and fine-grained austenite, which transforms into martensite on rapid quenching from the hardening temperature (Figure 11.7). The presence of martensite increases the hardness of the steel.

For hypoeutectoid steel if the hardening temperature is chosen in the same way as that of hypereutectoid steel (i.e., in between A_1 and A_3), a microstructure comprised of ferrite and austenite is formed. It transforms to ferrite and martensite on

Common Heat Treatment Practices

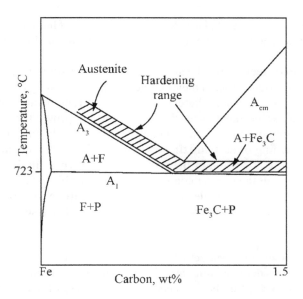

FIGURE 11.6 Hardening temperature ranges for different types of carbon steels. A, austenite; F, ferrite; P, pearlite.

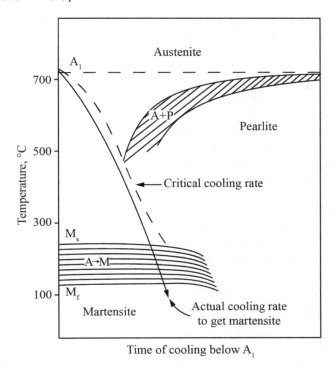

FIGURE 11.7 Critical cooling rate for martensite formation in eutectoid steel. A, austenite; M, martensite.

144   Phase Transformations and Heat Treatments of Steels

quenching. Ferrite, a very soft phase, lowers the hardness of hardened steel considerably. This phenomenon is known as incomplete hardening and is not usually used in practice.

Hypereutectoid steels, when heated at a temperature slightly above A_1, consist of fine grains of proeutectoid cementite and austenite. Upon quenching, this austenite changes to martensite, whereas the cementite remains unchanged. Hence, cementite being harder than martensite, its presence improves the hardness, abrasion resistance, and wear resistance as compared to the only martensite structure. Now, if the hardening or austenitizing temperature is increased and is just below the A_{cm} temperature, a portion of the proeutectoid cementite in the austenite gets dissolved to increase the carbon content of austenite. This excessively high temperature also tends to form undesirable coarse austenitic grain size. On quenching, the hardness of the as-quenched material is less because of the following reasons:

- A lesser amount of proeutectoid cementite is present.
- A more considerable amount of soft retained austenite is produced since the dissolved carbon of cementite has lowered both the M_s and M_f temperatures.
- The coarser martensite has a lesser hardness.

Furthermore, heating hypereutectoid steels above A_{cm} results in the austenite of coarser grains and surface decarburization. The as-quenched hardness is again low in this case because of the following reasons:

- Lack of harder cementite.
- A higher amount of dissolved carbon in the austenite results in a higher retained austenite content.
- Coarse acicular martensite has poor mechanical properties.
- Decarburized surface responds poorly to hardening treatment.
- Very high magnitude of internal stresses is introduced into the hardened structure as a result of quenching from a high temperature.

In the austenitic type of alloy steels, the room temperature structure may consist of austenite and some precipitate alloy carbides. These steels have M_s temperatures below the room temperature, and thus, heating these steels to temperatures higher than A_{cm} helps to dissolve the alloy carbides completely to obtain homogenous austenite.

In carbide type of alloy steels, heating is done at higher temperatures, i.e., between A_1 and A_{cm} to dissolve as much of alloy carbides as possible, leaving some carbides (such as vanadium carbide) in the undissolved state to inhibit the grain growth. The dissolved carbides are made to precipitate as fine and uniformly dispersed carbides during tempering, which induces in such steels the important characteristic called red hardness or secondary hardening. More is the volume of fine carbide precipitates; better is the red hardness. It makes these tool steels to cut or machine materials at high speeds.

11.2.4 QUENCHING

Quenching is the process that determines the cooling rate to decide the properties of the hardened material like structure, hardness, and strength. Quenching may be done by air, water, brine, oils, polymer quenchant, and salts or with the help of gases, liquids, and solids. Usually, a quenching medium is selected that provides a faster cooling rate than the critical cooling rate. Quenching medium characteristics are as follows:

- Quenchant temperature
- Specific heat capacity of the quenchant
- The viscosity of the quenchant
- Thermal conductivity
- Latent heat of vaporization
- Extent of agitation of the quenching medium

When a heated (say at 840°C) steel specimen is quenched in a stationary bath of cold water, a typical cooling curve forms with three stages (Figure 11.8).

- **Vapor blanket stage (stage A)**
  The surface of the hot metal gets vaporized as soon as the metal comes in contact with the quenching medium because of the high temperature of the metal. Hence, a thin stable film of vapor forms around the hot metal (Figure 11.9a). Since vapor films are poor conductors of heats, the cooling rate is relatively slow. At this stage, cooling is done by conduction and radiation through the vapor film.
- **Liquid boiling stage (stage B)**
  In this stage, heat is removed very rapidly in the form of heat of vaporization, and so the vapor blanket is broken, as shown in Figure 11.9b. After that, the quenchant comes directly in contact with hot metal surface, and violent boiling occurs. Very rapid cooling, mostly by convection, occurs in this stage as the quenching medium always remains in contact with the steel surface. This stage of quenching corresponds to the nose of the continuous cooling temperature (CCT) curve of the steel, i.e., 500°C–100°C.

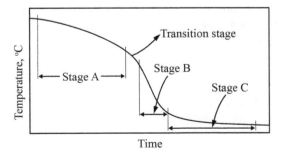

FIGURE 11.8 Cooling curve of a heated steel object when quenched in cold water.

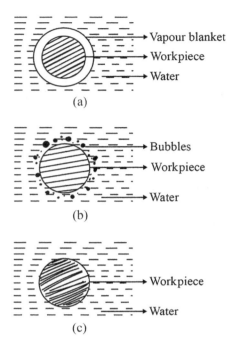

FIGURE 11.9 Schematic illustration of different stages of cooling: (a) vapor blanket stage, (b) liquid boiling stage, and (c) direct contact stage.

- **Direct contact stage (stage C)**

    It is also known as the liquid cooling stage or convection stage. It begins when the temperature of the metal surface is below the boiling point of the quenching liquid (Figure 11.9c). Cooling in this stage occurs by both conduction and convection processes through the liquid. The cooling rate is slowest in this stage as vapors do not form.

11.2.4.1 Quenching Media

Given the discussions about heat removal, which includes three stages, there are practically no such quenching media that can follow these three stages ideally. Let us discuss some of the industrially accepted quenching media in detail. Before we begin, the industrial quenching media in order of decreasing quenching severity are listed in the following:

- Brine
- Water
- Fused or liquid salts
- Soluble oil and water solutions
- Oil
- Air

Common Heat Treatment Practices

11.2.4.1.1 Brine

Aqueous solutions of salts like 10% sodium chloride or calcium chloride are denoted as brine solutions. In brine, due to the heating of the quenching medium in contact with the metal surface, crystals of the salt tend to get deposited on the metal surface. The layer of solid crystals so formed disrupts with minor explosive violence and chucks off a cloud of crystals. The vapor film gets destroyed, which allows direct contact of the brine with the metal surface. Brines are used where faster cooling rates are required.

11.2.4.1.2 Water

It is the most popular quenching medium because of its low cost, availability in abundance, secure handling, no pollution problems, and maximum cooling rate. Water quenching is only applicable for plain carbon steels and some low-alloy steels. Water quenching often results in a cooling rate, which is greater than the critical cooling rate. Such a high cooling rate usually leads to the creation of large amount of internal stresses in the quenched product, leading to distortion or even to the advance of cracks. Long-term stability of the vapor blanket stage is another noteworthy drawback of water quenching. Agitation of the water bath during quenching may help in reduction or elimination of this stage.

11.2.4.1.3 Salt Bath

Salt bath is especially used as a quenchant for tool steels. It is the best-suited quenching medium for steel with good hardenability and thin sections. Its advantages are a uniform temperature throughout the bath, selective hardening, no danger of oxidation, carburization, or decarburization during cooling. Some common salt baths are $NaNO_3$, 5% $NaNO_3$+50% KNO_3, and 50% $NaNO_3$+50% KNO_2.

11.2.4.1.4 Oil

Oils have high boiling points, lower quenching power than water or brine, lower stability, and high cost. Mineral oil is the most common oil used as quenchant.

11.2.4.2 As-Quenched Structures

At first, martensite forms as soon as the quenching starts. The depth of martensite formation depends on the time the part remains in the quenching bath. In quenched steel, the quantity of martensite formed depends on the location of M_s, M_f, and the temperature of the quenching medium, which usually is the room temperature (Figure 11.10). As long as the room temperature lies between M_s and M_f temperatures, austenite does not change completely to martensite since it has not been cooled to below the M_f temperature. The untransformed austenite is called retained austenite. The amount of retained austenite rises with the increase of carbon content of steel, as shown in Figure 11.11.

The substructure of the retained austenite consists of a higher density of dislocations and stacking faults.

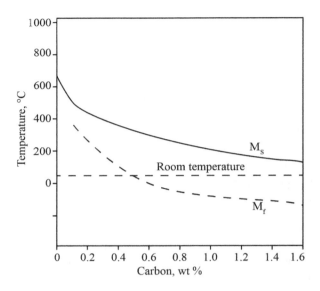

FIGURE 11.10 Influence of carbon on the M_s and M_f temperatures.

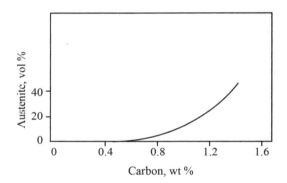

FIGURE 11.11 Influence of carbon on the amount of retained austenite.

11.2.5 Tempering

Tempering consists of heating the hardened steel to a temperature below the A_1 temperature, holding at that temperature, and then followed by prolonged cooling. The aim of tempering is to relieve quenching stresses developed during hardening and increase ductility and toughness of the steel by sacrificing the hardness or strength.

11.2.5.1 Stages of Tempering

Based on the tempering temperature, the treatment proceeds through four distinct but then again overlapping stages.

Common Heat Treatment Practices

11.2.5.1.1 First Stage of Tampering (up to 200°C)

During the early stage of tempering, precipitation of a hexagonal close-packed transition carbide (epsilon carbide, $Fe_{2.4}C$) occurs due to a decrease in the tetragonality of martensite. The formation of epsilon carbide decreases the c/a ratio as carbon precipitates from martensite.

Steels having carbon up to 0.2%, as a rule, are not hardened and tempered. However, martensite in such steels has a BCC structure, which does not show any change in the first stage of tempering except carbon atoms that segregate to dislocations. Steels having more than 0.2% carbon are highly unstable due to supersaturation. So, in this case, tempering results in the formation of low-carbon martensite and epsilon carbide by the transformation of high-carbon martensite. The contraction in volume occurs due to the rejection of carbon. Decrease of tetragonality decreases the hardness, but precipitation of epsilon carbide increases the hardness of the steel proportional to its amount formed. Thus, the net effect is that the hardness of steel usually decreases continuously but only slightly. In high-carbon steels (1.2% C steel), a slight increase in hardness is observed due to a relatively large volume of epsilon carbide.

11.2.5.1.2 Second Stage of Tempering (200°C–300°C)

During this stage, retained austenite transforms to lower bainite. This lower bainite consists of ferrite and epsilon carbide and, therefore, is different from the conventional bainite. A slight increase in the volume of steel occurs. When the carbon content of the steel is high, the amount of retained austenite being large transforms to more hard lower bainite in vast proportions. Ductility and toughness increase slightly by this treatment, with a corresponding decrease in hardness and strength.

11.2.5.1.3 Third Stage of Tempering (200°C–350°C)

In the third stage, the epsilon carbides dissolve, and the low-carbon martensite loses both its carbon and tetragonality and becomes ferrite. Carbon thus released combines with epsilon carbide, which in turn transforms to rod-shaped cementite. All these changes occur with the help of diffusion and nucleation. The contraction of volume occurs at this stage. The hardness decreases continuously and sharply with a rapid increase in toughness.

11.2.5.1.4 Fourth Stage of Tempering (350°C–700°C)

Heating the hardened steel in this range of temperature results in coarsening or spheroidization of cementite along with recovery and recrystallization of ferrite. The growth of cementite starts at around 300°C, but spheroidization occurs above 400°C. Reduction in surface energy provides the necessary condition for spheroidization of cementite. Above 600°C, equiaxed grains of ferrite (by recrystallization) form have coarse globules of cementite. It is the spheroidized or globular pearlite, which is softest with the highest ductility and best machinability.

11.2.5.1.5 Secondary Hardening and the Fifth Stage of Tempering

When alloying elements are added to steels, they may enter the ferrite or the carbides in varying amounts depending on the alloying element concerned. Non–carbide-forming elements, such as Al, Cu, Si, P, Ni, and Zr, enter into ferrite. They have minimal impact on the tempered hardness of steel. Other elements are found in both the ferrite and the carbides. A number of these elements in the order of their tendency to form carbides (Mn having the least and Ti the greatest) are Mn, Cr, W, Mo, V, and Ti. Most alloying elements in steels tend to increase the resistance of the steel to softening when it is heated, which means that for a given time and temperature of tempering, alloy steel will possess a higher hardness after tempering than plain carbon steel of the same carbon content. For example, there is an increase in hardness with the addition of 5% Mo in a 0.35% carbon steel (Figure 11.12). This effect is especially significant in steels that contain considerable amounts of carbide-forming elements. When the carbide-forming elements are tempered below 540°C, the tempering reactions tend to form cementite particles based on Fe_3C, or more accurately $(Fe, M)_3C$ where M represents any of the substitutional atoms in the steel. In general, the alloying elements are present in the cementite particles only in about the same ratio as they are present in the steel as a whole. When the tempering temperatures exceed 540°C, appreciable amounts of alloy carbides are precipitated, which replace the coarse cementite particles. The precipitation of alloy carbides is identified as the fifth stage of tempering. The precipitation of these new carbides, in general, does not conform to the formula $(Fe, M)_3C$.

11.2.5.1.6 Secondary Hardening

When the amount of strong carbide-forming elements is large, and the highly alloyed steels are tempered at 500°C–600°C, the general decrease of hardness with tempering temperature is not only arrested, but an increase in hardness with improved toughness occurs. This increase in hardness is called secondary hardness or red hardness. Above 500°C, these elements have high diffusivity to nucleate and grow

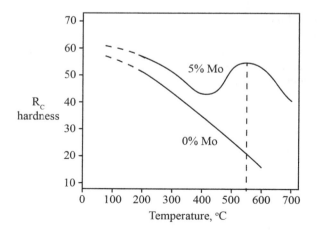

FIGURE 11.12 Secondary hardening caused by the presence of a 5% molybdenum in plain carbon steel.

Common Heat Treatment Practices

to form fine dispersion of alloy carbides to cause secondary hardening. Because of the lower diffusion rates of alloying elements, these alloy carbides are formed only at the high temperatures. Besides, at low temperatures, diffusion of solute atoms is sluggish to result in formation of any carbides.

Secondary hardening is a process in which coarse cementite particles are replaced by new and much finer alloy carbide dispersion of V_4C_3, Mo_2C, and W_2C. The critical dispersion causes a peak in the hardness, but as carbide dispersion slowly coarsens, the hardness decreases. The peak is realized at 500°C in chromium steels, 550°C in molybdenum steels, 550°C–600°C in vanadium steels, and 600°C in titanium steels. The amount of secondary hardening is directly proportional to the volume fraction of alloy carbides and, hence, is related to the concentration of the strong carbide-forming elements present in the steel.

11.2.6 Austempering and Martempering

Austempering is a hardening treatment in which austenite transforms isothermally to lower bainite and is used to reduce distortion and cracks in high-carbon steels. It involves heating the steel to above the austenitizing temperature, quenching in molten salt bath held at a particular temperature above M_s point and within the bainitic range (300°C–400°C), and then keeping the steel at this temperature to let austenite transform completely to lower bainite. After full transformation, steel is removed out of the bath and cooled in the air up to the room temperature. A schematic illustration of the austempering process superimposed on the TTT diagram is shown in Figure 11.13. The equalization of temperature throughout the cross section of the part before bainite formation minimizes the stresses developed during austempering,

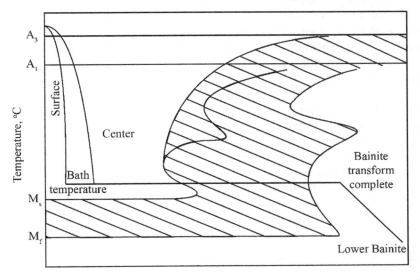

FIGURE 11.13 Schematic illustration of the austempering process.

152 Phase Transformations and Heat Treatments of Steels

which are negligible compared to stresses developed during hardening. Steel to be austempered should have adequate hardenability to avoid pearlite formation. TTT diagram is also useful for determining the suitability of given steel for austempering. The rate of cooling should be higher than the upper critical cooling rate, i.e., care must be taken such that there is no austenite to pearlite transformation. Steels having considerable soaking time for the bainite transformation are not suitable for this treatment.

Martempering is also a hardening treatment used to minimize distortion and cracking. In regular hardening operation, when the heated steel is plunged in coolant, a vast difference of temperature develops between the surface and the center of the part, which introduces two types of stresses:

11.2.6.1 Thermal Stresses

A significant difference in temperature results in differential volume contraction, which produces volumetric internal stresses called thermal stresses.

11.2.6.2 Structural Stresses

Due to a significant difference in temperature between the surface and the center, the martensite transformation range is attained at a different time from the surface to the center. This differential expansion causes internal stresses called structural stresses.

When both these types of stresses develop simultaneously, the nature of stress becomes quite complicated at places, and therefore, the magnitude of stresses may become large enough to cause distortion and even cracking. If the magnitude of the tensile stress exceeds the yield point of the material, the steel distorts, but if it exceeds the tensile strength, the steel develops cracks.

Martempering treatment involves heating the steel to the austenitizing temperature followed by quenching in a constant temperature bath maintained above M_s point, typically between 180°C and 250°C. Steel is then held in the bath until the temperature throughout the cross section (from the surface to the center) becomes uniform and identical to the bath temperature. After that, steel is withdrawn and cooled in air through M_s to room temperature to obtain martensite simultaneously across the whole section. Finally, tempering is done as required for the part. Figure 11.14 explains the martempering process superimposed on the TTT diagram. The cooling rate should be substantially high, and the soaking period quite short to restrict the conversion of austenite to pearlite or even bainite. In martempering, as the bath temperature is not the room temperature but above M_s temperature, the steel temperature gradient is reduced, and so, the magnitude of the thermal stress is reduced. TTT is useful to fix the correct bath temperature, to fix the isothermal holding time without bainite formation, and to find out the maximum size of the part of that steel, which can be martempered.

11.2.7 Subzero Treatment

Subzero treatment is mainly done for hardened steels that contain retained austenite. As discussed in the quenching section, retained austenite drastically reduces the mechanical properties and leads to unpredictable dimensional changes, and

Common Heat Treatment Practices

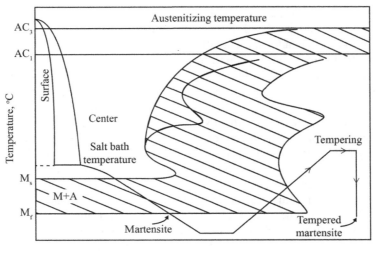

FIGURE 11.14 Schematic illustration of the martempering process.

therefore, treatment is required to convert them to martensite, which is done by the subzero treatment. The process involves cooling the steel to the subzero temperature that is lower than the M_f temperature of the steel. Usually, M_f temperatures for steels lie between −30°C and −70°C. This treatment is followed by tempering to remove the internal stresses developed in the steel. The subzero treatment is not useful for such steels that are slightly heated to 100°C–200°C or kept at room temperature for several hours after hardening because austenite becomes stable, and then it does not transform to martensite completely. Hence, subzero treatment should always be executed immediately after completion of hardening treatment. Subzero treatment is mostly preferred in high-carbon and high-alloy steels, making them suitable for bearings, gauges, tools, and components that require high impact and fatigue strength in addition to dimensional stability.

11.3 HARDENABILITY

11.3.1 Significance

Steels are typically graded based on chemical composition. However, the composition is not a fixed number; instead, it is a range. For example, AISI 4340 steel has 0.38%–0.43% carbon, 0.60%–0.80% manganese, 0.20%–0.35% silicon, 1.65%–2.00% nickel, 0.70%–0.90% chromium, and 0.20%–0.30% molybdenum. This variation in the chemical composition causes a difference in the critical cooling rate during quenching. Hence, the rates of cooling at the surface and center are not the same, which in turn results in a gradient of hardness across the cross section. Purchasing steel based on chemical composition will not guarantee the full hardness, and therefore, it is necessary to know the hardenability of the steel.

Hardenability may be defined as the susceptibility of the steel to hardening when quenched and is related to the depth and distribution of hardness across a cross section. It depends on the addition of alloying elements and the grain size of austenite. If rapid cooling is performed, martensitic transformation may even be feasible at the center of the steel if the corresponding cooling rate exceeds the critical one. In this way, a completely hardened steel would be obtained with martensite throughout the cross section. However, such rapid quenching may further be associated with unwanted effects like warping and ultimately cracking of the specimen. Therefore, the hardenability of steel is revealed by its ability to harden throughout its cross section while avoiding radical quenching. The regions of martensite and pearlite can be easily differentiated in the steel section in many ways. The presence of pearlite results in a sudden fall of hardness. The fracture surface of the sample, as schematically illustrated in Figure 11.15, is another way of identifying the martensite or pearlite content. While martensite breaks with brittle fracture, pearlite breaks with a ductile fracture. Microstructurally, steel contains 50% pearlite and 50% martensite at the brittle-to-ductile transition region.

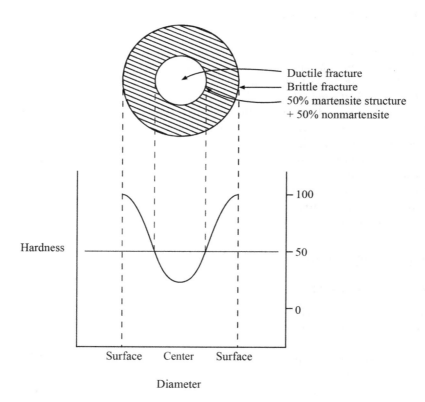

FIGURE 11.15 Variation of hardness across the cross section of a quenched cylinder. Microstructural examination reveals that the fracture surface has two distinct appearances, i.e., ductile and brittle fracture.

Common Heat Treatment Practices

11.3.2 Determination of Hardenability

The following methods determine the hardenability of steel:

- Grossman's critical diameter
- Jominy end quench test

11.3.2.1 Grossman's Critical Diameter

The test is named after M.A. Grossman, who measured hardenability in terms of critical diameter. In this method, many steel rods of different diameters are quenched under the same conditions (Figure 11.16). While the rod with the smallest diameter gets hardened throughout the cross section, the hardness decreases in the center with an increase in the diameter of the rod. As the diameter of the rod increases, the cooling rate at the center of the steel rod decreases, thereby giving a softcore containing pearlite. The position of the steepest drop in hardness, or when the rod has a 50% martensitic structure at the center, is called the critical diameter (D_c). Its value depends on the steel composition and the quenching method. For example, the value of critical diameter is 1.83″ for water and 1.25″ for oil. It is always essential to specify the severity of the quench when hardenability is to be expressed in terms of D_c.

11.3.2.2 Severity of Quench

Grossman defined an ideal quenching medium as a medium that removes the heat from the steel surface as soon as the heat flows from the center of the rod to the surface. The critical diameter of the specimen corresponding to this ideal quenchant is termed as the ideal critical diameter (D_I). On the other hand, the severity of the quench is designated by the heat transfer equivalent H and is given by

$$H = \frac{S}{K} \qquad (11.2)$$

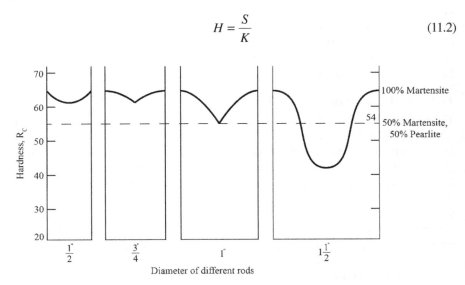

FIGURE 11.16 Effect of the diameter of the specimen on the depth or hardness while quenching in the same medium.

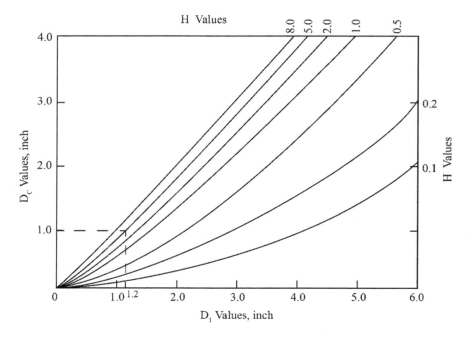

FIGURE 11.17 Grossman's master graph, which shows the relationship between the critical diameter (D_c), ideal critical diameter (D_I), and the severity of quench (H).

where S is the heat transfer coefficient between the steel and the quenching medium, and K is the thermal conductivity of the steel. The relationship between the critical diameter (D_c), ideal critical diameter (D_I), and the severity of quench (H) can be obtained from thermodynamic considerations, as shown in Grossman's master graph (Figure 11.17). From the figure, for $D_c = 1''$ and $H = 5$ (agitated brine quench), D_I or hardenability comes out to be $1.2''$.

11.3.2.3 Jominy End Quench Test

The Grossman method of determining the hardenability of steel is not so popular as it takes a long time to quench and test a large number of steel rods. A comparatively faster and popular method of hardenability determination of steel is the Jominy end quench test. In this test, a round specimen having a diameter of $1''$ and a length of $4''$ is heated uniformly to the proper austenitizing temperature (Figure 11.18a). After that, the sample is quickly placed in a fixture where a jet of water impinges the bottom part of the specimen. Specific parameters such as the size of the orifice, the distance from the orifice to the bottom of the specimen, and the temperature and circulation of the water are standardized so that every specimen in this fixture can receive the same rate of cooling. After cooling, the specimen is removed from the fixture, and two parallel flat surfaces are ground longitudinally to a depth of $0.015''$. The hardness is obtained at intervals of $1/16''$ from the quenched end. A curve is then drawn between the hardness values and the distance from the quenched end

Common Heat Treatment Practices

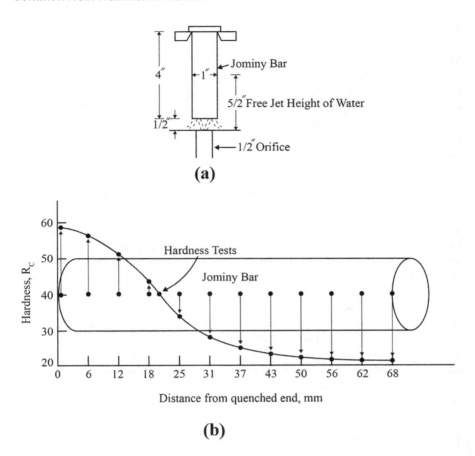

FIGURE 11.18 Schematic illustration of the Jominy end quench test: (a) experimental setup and (b) a typical Jominy curve showing the hardness variations in the quenched bar.

(Figure 11.18b). In general, ASTM Designation A225-48T End Quench Test is followed for determining the hardenability of steel.

11.3.3 Factors Affecting Hardenability

Hardenability is a measure of the amount of martensite present in the structure. Hence, steel is said to have high hardenability if it is completely transformed to martensite. The factors that promote the formation of martensite are those who shift the nose of the TTT curve to the right. Alloying elements, carbon content, and austenitic grain size affect hardenability. All alloying elements, except cobalt, have a tendency to shift the nose of the TTT curve to the right. This ensures a more hardenable steel as the probability that the cooling curve will not touch the nose of CCT curve is higher. However, undissolved inclusions, such as carbides or nitrides, nonmetallic inclusions, and inhomogeneity of austenite, which may happen due to the presence of alloying elements, decrease the hardenability of the steel. Carbon content dramatically influences the hardenability. The size of austenite also plays a vital role in determining the hardenability. The finer the austenite grain size, the lower the hardenability. It may be attributed to the fact that with a decrease in grain size, more sites become available for pearlite nucleation. Once pearlite formation occurs, martensite transformation is suppressed, and hence, hardenability decreases. Although coarse-grained steel increases the hardenability, it is not always recommended to use the same to increase hardenability because it may lead to specific other undesirable properties such as poor impact properties, decrease in ductility, and quench crack susceptibility.

11.4 CASE HARDENING AND SURFACE HARDENING

A large number of machine components like gears require a combination of various properties such as high surface hardness and wear resistance along with excellent toughness and impact resistance. High-carbon steels can provide high surface hardness and wear resistance after suitable heat treatment, but impact strength is poor. On the other hand, low-carbon steels provide high impact strengths but very low surface hardness and wear resistance. Surface hardening and case hardening methods are commonly used to impart this combination of properties. While case hardening changes the chemical composition of the surface layers, surface hardening does not alter the chemical composition of the steel but includes phase transformation by fast heating and cooling of the outer surface. Conventional case hardening methods are carburizing, nitriding, and cyaniding or carbonitriding. Surface hardening processes are flame hardening, induction hardening, laser hardening, and electron beam hardening. Let us discuss some of these methods in detail.

11.4.1 CARBURIZING

It is a heat treatment operation in which the carbon content of the surface of low-carbon steel, usually about 0.20% carbon or lower, or low-carbon-containing alloy steel is enriched by exposing the steel to a carburizing atmosphere in the austenitic region followed by quenching and subsequent tempering. Generally, carburizing is carried out by keeping the low-carbon steel in contact with solid, liquid, or gaseous atmosphere of high-carbon activity at high temperatures (850°C–950°C) to obtain surface carbon content of 0.6%–1.10%. Even with higher surface carbon content, the case becomes very brittle due to the coarse cementite network formed in it. So the steel is subjected to quenching. As-quenched structures formed are martensite and retained austenite. However, the surface is still brittle. Therefore, the steel is subsequently subjected to tempering.

11.4.1.1 Mechanism of Carburizing

Carburizing occurs in two steps:

- Absorption of free carbon on the steel surface occurs due to a large difference between the carbon potential of the atmosphere and the carbon content of the steel surface. The absorption rate of the additional carbon content on

Common Heat Treatment Practices

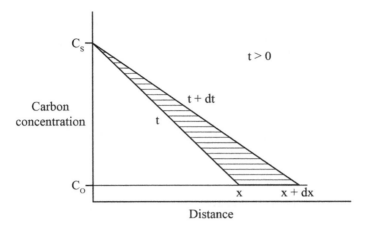

FIGURE 11.19 Variation of carbon concentration with distance.

the surface increases with an increase in the surface carbon content until it becomes equal to that of the atmosphere.

$$CO_2(g) + C(s) \rightleftharpoons 2CO(g) \tag{11.3}$$

- The diffusion of carbon from the surface to the interior occurs in the second step. As soon as the steel comes in contact with the carburizing atmosphere, the carbon on the surface attains its maximum value (C_s). It remains constant throughout the process. The carbon content of the interior is the same as the original carbon content in the steel, C_o. At an instant $t > 0$, the carbon concentration varies linearly with distance (Figure 11.19). Fick's first law of diffusion then governs the diffusion of carbon into the steel.

11.4.1.2 Fick's First Law of Diffusion

Fick's first law of diffusion is given by

$$J = -D_\gamma^c \frac{dc}{dx} = D_\gamma^c \frac{C_s - C_o}{dx} \tag{11.4}$$

where J is the flux of diffusing species. Here, J is the net carbon flux per unit area per second. D_γ^c is the diffusivity of carbon in austenite, and $\dfrac{dc}{dx}$ is the concentration gradient. The negative sign indicates that the concentration of carbon decreases with distance. Let us assume that the case depth increases from x to $x + dx$ on increasing the duration from t to $t + dt$. In the time interval dt, the carbon supplied by diffusion is Jdt, which is shown by the shaded area in Figure 11.19. So,

$$Jdt = \frac{(C_s - C_o)(x + dx)}{2} - \frac{(C_s - C_o)(x)}{2} \tag{11.5}$$

On combining equations 11.4 and 11.5, we have

$$D_\gamma^c \frac{(C_s - C_o)}{x} dt = \frac{(C_s - C_o)(x + dx) - (C_s - C_o)(x)}{2} \quad (11.6)$$

$$D_\gamma^c \frac{(C_s - C_o)}{x} dt = \frac{(C_s - C_o)(dx)}{2} \quad (11.7)$$

$$2D_\gamma^c (C_s - C_o) dt = (C_s - C_o) x(dx) \quad (11.8)$$

$$x dx = 2 D_\gamma^c dt \quad (11.9)$$

On integrating equation 11.9, we have

$$\frac{x^2}{2} = 2 D_\gamma^c t \quad (11.10)$$

$$x = \emptyset \sqrt{t} \quad (11.11)$$

where x is total case depth in milimeters, t is the time in hours, and $\emptyset$ parameter is a function of temperature, as illustrated in Figure 11.20.

11.4.1.3 Heat Treatment after Carburizing (Postcarburizing Treatment)

Carburized components are given heat treatment to develop hard and wear-resistant surfaces, to refine the grain size of the case as well as core, and to break the cementite network to reduce brittleness. Typically, a double heat treatment cycle is performed in case the steel is having a coarse-grained structure throughout its cross section. The steel is first subjected to above A_3 temperature to refine the grain size of the core and dissolve the proeutectoid cementite of the case. Normalizing is then done, which does not allow the network of cementite to form. It is then reheated to above AC_1 to refine the grain size of the case, and quenching produces fine martensite in

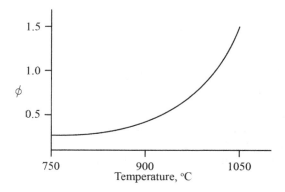

FIGURE 11.20 Variation of $\emptyset$ parameter with temperature.

Common Heat Treatment Practices

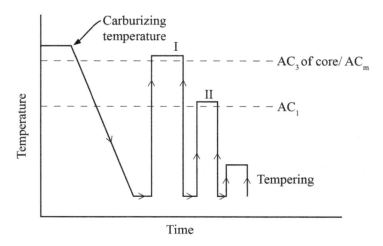

FIGURE 11.21 Schematic representation of postcarburizing heat treatment cycles.

the case. Tempering is done at 180°C–200°C to relieve the stresses and induce some toughness. A schematic representation of this treatment is provided in Figure 11.21.

11.4.2 Nitriding

Nitriding is another case hardening technique where nitrogen enrichment of the surface of the steel specimen is done in the temperature range of 500°C–575°C (when the steel is ferritic) in an atmosphere of 15%–30% dissociated ammonia for an extended time. Nitriding does not require heating of the steel into austenitic field region and subsequent quenching to form martensite. So, it involves minimum distortion and provides excellent dimensional control. Steels subjected to nitriding are medium carbon steels. Main reasons for the high surface hardness of nitrided steels are as follows:

- Fine (size 5–15 nm) and uniformly dispersed particles act as strong barriers to block the motions of dislocations.
- There is a high density of dislocations of $10^{10} cm^{-2}$ as in a heavily cold-worked metal, which increases the hardness.
- Nitrogen atoms form Cottrell-type atmospheres to increase the hardness.

11.4.2.1 Gas Nitriding

Gas nitriding is the most common method of nitriding, where the components to be nitrided are heated in an atmosphere of dissociated ammonia gas. At nitriding temperatures, 500°C–575°C, ammonia reacts at the surface of the steel to introduce nitrogen in iron, i.e.;

$$NH_3 = N(Fe) + \frac{3}{2}H_2 \qquad (11.12)$$

162 Phase Transformations and Heat Treatments of Steels

with a constant equilibrium K given by

$$K_1 = \frac{a_N p_{H_2}^{3/2}}{p_{NH_3}}$$ (11.13)

$$\%N = \frac{K_1 p_{NH_3}}{p_{H_2}^{3/2}}$$ (11.14)

$\%N$ refers to nitrogen dissolved in iron at a given temperature. During nitriding, $\%N$ at the surface in solution is given by equation 11.14, only when it does not exceed the solubility limit of nitrogen in steel at the nitriding temperature. Ammonia gas is itself unstable at nitriding temperatures and dissociates according to the reaction, as given in the following:

$$NH_3 = \frac{1}{2}N_2 + \frac{3}{2}H_2$$ (11.15)

This reaction is relatively slow. Molecular nitrogen present in the gas atmosphere may also decompose at the surface of the steel to introduce nitrogen in steel, i.e.,

$$\frac{1}{2}N_2 = N(Fe)$$ (11.16)

$$\%N = K_2 p_{N_2}^{1/2}$$ (11.17)

However, at nitriding temperatures, the constant K_2 is orders of magnitude smaller than K_1 in equation 11.13. NH_3-H_2 mixture containing 18% NH_3 and 82% H_2 (total pressure of 1 atm) gives the same percentage nitrogen dissolved in iron as 5×10^3 atm of N_2 at 500°C. Hence, it is not practical to use nitrogen gas for nitriding; instead, we go for $NH_3 + N_2 + H_2$ mixture. The atomic nitrogen in equation 11.12, which is absorbed by the steel surface, can form nitrides following the phase diagram of the iron–nitrogen system. The first phase is $\alpha_{ss} - N$, which is called nitrous ferrite. Another vital phase that forms is α'' (low concentration of N). It is a fine and coherent precipitate ($Fe_{16}N_2$). In case the level of nitrogen exceeds 0.1%, γ'-nitride forms, which is Fe_4N. This γ'-nitride contributes to the brittle white layer of nitride on steel surface. If nitrogen >6%, then ϵ-nitride forms, which is a solid solution, and when it is combined with carbon, it produces a tribologically desirable phase. The duration of nitriding usually varies from 48 to 96h to achieve 1-mm-deep layer. It also depends upon the temperature (Figure 11.22), alloying elements, and the depth required.

11.4.3 CYANIDING AND CARBONITRIDING

The cyaniding and carbonitriding processes involve the hardening of the surface layer of the steel by the addition of both carbon and nitrogen. If the hardening is carried out in liquid salt baths, it is known as cyaniding, and if the process is carried out in a gaseous atmosphere, it is called as carbonitriding.

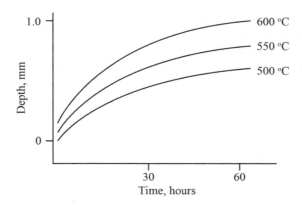

FIGURE 11.22 Effect of nitriding time and temperature on the case depth.

In the cyaniding process, the parts to be hardened are immersed in a liquid bath of NaCN with the concentration changing between 25% and 90% at 800°C–960°C. A controlled flow of air is passed through the molten bath. Molten NaCN decomposes in the presence of air to produce sodium cyanate (NaNCO), which in turn decomposes to produce carbon and nitrogen. The reactions involved are as follows:

$$2NaCN + O_2 \rightarrow 2NaNCO \tag{11.18}$$

$$2NaNCO + O_2 \rightarrow Na_2CO_3 + CO + 2N \tag{11.19}$$

$$2CO \rightarrow CO_2 + C \tag{11.20}$$

The carbon content of the surface increases with an increase in the cyanide concentration of the bath. Generally, the hardened parts are quenched in water or oil after cyaniding, followed by a low-temperature tempering.

The carbonitriding process involves the hardening of the steel in a gas mixture of carrier gas, enriching gas, and ammonia so that carbon and nitrogen are diffused simultaneously into the surface of the steel. While the carrier gas is typically a mixture of nitrogen, hydrogen, and carbon monoxide produced in an endothermic generator, as in gas carburizing, the enriching gas is usually propane or natural gas and is the primary source for the carbon added to the surface. Nitrogen content diffused into the steel surface depends on the ammonia content and temperature. This process can be carried out at a lower temperature since carbon–nitrogen austenite is stable at lower temperatures than plain carbon austenite. Here, quenching is performed by oil rather than water to avoid distortion and cracking.

11.4.4 Flame Hardening

In flame hardening, heat is applied by an oxy-acetylene or oxy-fuel blowpipe, followed by spraying of a jet of water as the coolant. Skill is required to avoid overheating; otherwise, it can result in cracking after quenching and grain growth in

164 Phase Transformations and Heat Treatments of Steels

the areas just below the hardened surface. Since selected areas are hardened in this treatment and then quenched to form martensite, it is essential to begin with a steel that is capable of being hardened. Typically, the carbon content of steel used for flame hardening varies from 0.3% to 0.6%. In general, four methods are used for flame hardening:

- Stationary
- Progressive
- Spinning
- Progressive–spinning

In the stationary method, both torch and workpiece are stationary, and therefore, this method is used for the spot hardening of small parts such as valve stems and open-end wrenches. In the progressive method, the torch moves over the large stationary workpiece. This method is utilized for the hardening of large parts like teeth of giant gears. In the spinning method, the torch is fixed while the workpiece rotates. This method is used for circular cross section, such as precision gears and pulleys. In the progressive spinning method, the torch moves over the rotating workpiece, and the method is used to surface-harden large circular parts such as rolls and shafts.

11.4.5 INDUCTION HARDENING

As the name suggests, this treatment involves heating of the components by electromagnetic induction. A coil or conductor carries a high-frequency alternating current, which is then induced into the steel component enclosed within the magnetic field of the coil. As a result, induction heating takes place, and only the outer surface of the steel component is hardened due to the skin effect. However, the heat applied to the surface tends to flow toward the center of the component by conduction, and therefore, the time of hardening is an essential parameter in controlling the depth of the hardened zone. Typically, the parts are heated for a few seconds only to a depth that is inversely proportional to the square root of the frequency followed by quenching by a jet of water. Frequencies in the range of 10,000–500,000 Hz are commonly used in this hardening method.

11.5 THERMOMECHANICAL TREATMENT

Thermomechanical treatment is a combination of heat treatment and a deformation process to change the shape and refine the microstructure of the material by affecting the phase transformation. The mechanism behind this treatment is that the plastic deformation process induces various crystal defects such as dislocations, vacancies, stacking faults, and subgrain boundaries in the material. These defect sites act as nucleation sites and thereby alter the kinetics of phase transformation and morphology of the phases developed in the specimen. Hot rolling of metals, a well-established industrial process, is a thermomechanical treatment that plays an integral part in the processing of many sheets of steel from low-carbon mild steels to highly alloyed stainless steels.

Common Heat Treatment Practices 165

11.5.1 CLASSIFICATION BASED ON CRITICAL TEMPERATURES

Table 11.1 gives the critical temperatures–based classification of thermomechanical treatments of steel. Three classes of thermomechanical treatments have been identified, namely, supercritical, intercritical, and subcritical treatment. While supercritical treatment involves deforming the steel in the austenitic condition, the intercritical treatment comprises deforming the steel in a phase mixture of ferrite and austenite. On the other hand, subcritical treatment comprises deforming the steel below the lower critical temperature (A_1). This treatment also comes under the low-temperature thermomechanical treatment (LTMT) since the austenite is supercooled to a temperature below the recrystallization temperature. Accordingly, the supercritical treatment that includes deformation of austenite at a temperature above the recrystallization temperature falls under the high-temperature thermomechanical treatment category. This section mostly focuses on the subcritical treatment or LTMT such as ausforming and isoforming processes.

11.5.1.1 Ausforming

The process involves the deformation of metastable austenite at a constant temperature between the ferrite and bainite curves of the TTT diagram, as shown in Figure 11.23. No transformation occurs during mechanical treatment. The deformed steel is then quenched to obtain martensite, and subsequent tempering is done to achieve appropriate mechanical properties. It provides high strength without affecting ductility and toughness. Usually, a 4.7% Cr, 1.5% Mo, 0.4% V, and 0.34% C steel has a tensile strength of about 2000 MPa after conventional quenching and tempering treatment, whereas after ausforming, the strength increases to over 3000 MPa. Steels that possess a sufficient gap between the pearlite and bainitic curves are subjected to this treatment. Alloying elements such as Cr, Mo, Ni, and Mn develop a deep metastable austenite bay by displacing the TTT curve to longer transformation times (Figure 11.23). In this case, a high austenitization temperature is necessary for the dissolution of alloying elements. Nevertheless, it should not be very high; otherwise, the coarsening of austenite will take place, which ultimately results in coarse martensite (low toughness) on quenching. Cooling from this austenitizing temperature to the metastable austenite bay must be rapid to avoid the occurrence of ferrite, and after deformation, the cooling must be fast enough to prevent the formation of bainite. The amount of deformation is the most critical variable. The higher

TABLE 11.1
Critical Temperature–Based Classification of Thermomechanical Treatments

Class	Temperature Range	Examples
Supercritical TMT or HTMT	Above A_3	Hot–cold working, controlled rolling
Intercritical TMT	Between A_3 and A_1	
Subcritical TMT or LTMT	Below A_1	Isoforming, ausforming

TMT, thermomechanical treatment; LTMT, low-temperature thermomechanical treatment; HTMT, high-temperature thermomechanical treatment.

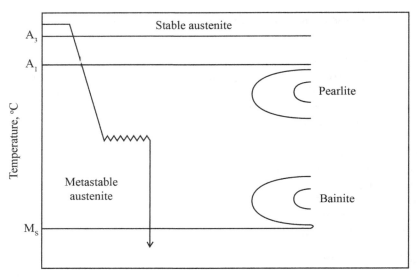

FIGURE 11.23 Schematic illustration of the ausforming process.

the degree of working, the greater the ultimate strength achieved. Ductility increases with increasing deformation, although it only becomes significant at deformations above 30% reduction in thickness. Steels subjected to substantial deformation during ausforming process exhibit very high dislocation densities (up to 10^{13} cm^{-2}). This high dislocation density along with the fine dispersion of alloy carbides is responsible for the strengthening effect.

11.5.1.2 Isoforming

Unlike ausforming, isoforming involves deformation of metastable austenite until the transformation of austenite is complete at the deformation temperature. The steel is first heated to above A_3 followed by immediate quenching to a temperature of about 650°C, as shown in Figure 11.24. Mechanical deformation is applied at this temperature during the phase transformation so that the structure produced comprises fine ferrite subgrains with spheroidized cementite particles instead of a ferrite/pearlite aggregate. Due to the refinement of the ferrite grain size and replacement of lamellar cementite by spheroidized particles, substantial improvement in toughness and strength occurs.

11.6 HEAT TREATMENT OF CARBON AND ALLOY STEELS

In general, carbon steels are used in applications where moderate strength is required. They have the inherent properties derived from carbon only. Heat treatment of carbon steels is only possible in thinner sections, as it is challenging to harden thicker sections (thicker than 1.5 cm) because of their low hardenability. Moreover, these carbon steels cannot be used in corrosive environments. These limitations put engineers

Common Heat Treatment Practices

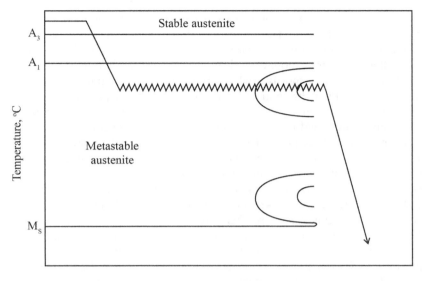

FIGURE 11.24 Schematic illustration of the isoforming process.

to think about alloy steels, which not only enhances the properties of carbon steels but also induces other specific properties in the steels. Heat treatment of carbon steels has been mostly discussed while describing standard heat treatment practices. Hence, this section focuses on various types of alloy steels and their heat treatment methods. However, emphasis on special steels and their heat treatment practices is deliberated in the subsequent chapter.

Alloy steels are categorized based on the alloying element present in the steel. The most common are manganese steels, silicon steels, chromium steels, nickel steels, and molybdenum steels. Manganese is added to carbon steels to improve the tensile strength, hardenability, and hot workability of steel. Heat treatment methods like hardening and tempering result in the best possible mechanical properties, while normalizing treatment improves the impact strength, after which they are used for large forgings and castings. Silicon, like manganese, is present as a cheap deoxidizer in all steels. When the steels contain more than 0.6% silicon, they are grouped as silicon steels. They possess improved elastic properties, excellent electrical and magnetic properties, and enhanced resistance to scaling at high temperatures. Mainly, steels with 3%–4% silicon and less than 0.5% carbon, popularly known as electrical steel, is used in the cores and poles of electrical machinery. The desired properties are derived from a coarse-grained and textured structure, which is obtained by repeated cold rolling and annealing at 1100°C–1200°C under hydrogen atmosphere.

On the other hand, chromium steels are well known for high hardness and wear resistance. The low-carbon (<0.2% carbon), low-chromium (<1% chromium) steels are usually case carburized. The presence of chromium increases the wear resistance of the case, but the toughness of the core is reduced. Hence, these low-carbon alloy steels are often subjected to postcarburizing heat treatment processes to refine both the case and the core. With an increase in carbon content up to 0.4%, these medium-carbon alloy steels are exposed to oil or water quenching, followed by tempering. These types of steels are mostly used in springs, engine bolts, axles, and studs. High-carbon (~1%) and high-chromium (~1.5%) alloy steels find application extensively in the ball and roller bearings and for crushing machinery. Nickel steels are best suited where high strength and toughness, improved fatigue strength, impact resistance, and shear strength are required, like for high-strength structural steels. These alloy steels, unlike carbon steels, do not need rapid quenching for obtaining high hardness since they are heat-treated at much lower temperatures. Low-carbon (~0.2%) and low-nickel (~3.5%) steels, typically from 23xx series, are widely used for carburizing of drive gears, studs, and kingpins, whereas the 5% nickel steels (25xx series) have improved toughness and are, therefore, used for heavy-duty applications such as bus and truck gears, cams, and crankshafts. Molybdenum steels require high heat treatment temperatures, typically 20°C higher than carbon steels of same carbon content, as they form complex carbides that are more stable than cementite. The alloy steels with low-carbon content (40xx and 44xx series) are generally used for case carburizing. With higher carbon content, they find applications in automotive coil and leaf springs.

Nevertheless, alloy steels with more than one alloying element, such as chromium–molybdenum steels, nickel–molybdenum steels, and nickel–chromium steels, have the added advantage of improved properties and are widely used in various sectors. Since a large number of such alloy steels are available, it is not possible to describe all the alloy steels types; however, a detailed description of special steels, which have unique applications, is given in the next chapter.

FURTHER READING

Singh, V. *Heat Treatment of Metals*. (Standard Publishers Distributors, New Delhi, 2006).
Rajan, T. V., Sharma, C. P., & Sharma, A. *Heat Treatment: Principles and Techniques*. (PHI Learning Pvt. Ltd., New Delhi, 2011).
Avala, L. *Concepts in Physical Metallurgy*. (Morgan & Claypool Publishers, San Rafael, CA, 2017).

12 Special Steels

12.1 STAINLESS STEELS

Stainless steels have a minimum of 11.5% chromium and some amounts of nickel, molybdenum, and manganese to enhance corrosion resistance. The corrosion-resisting property is due to the thin, protective, and stable chromium or nickel oxide film on the surface. A three-digit numbering system is used to identify stainless steel, as given in Table 12.1. While the first digit designates the group, the last two digits have no specific importance.

We have already discussed the fact that chromium is a suitable ferrite stabilizer, and hence, in the Fe–Cr phase diagram (Figure 12.1), it can be seen that chromium stabilizes ferrite, leading to the formation of a closed γ-loop with a maximum of 12.7% chromium at around 1000°C. The ferrite phase is stable above this chromium content of 12.7% over the entire temperature range. However, certain additions of austenite stabilizers such as carbon or nitrogen increase the solubility of chromium in austenite by expanding the γ-loop. Similarly, nickel further modifies the phase diagram, as it strongly stabilizes austenite. Figure 12.2 explains the effect of carbon on the phase diagram of 18/8 steel (18% chromium and 8% nickel). For a low-carbon 18/8 steel, carbides like $Cr_{23}C_6$ can be obtained at room temperatures. This carbide dissolves in austenite if the steel is heated to above 1050°C, and quenching from this temperature will avoid the formation of carbide and result in only austenite at room temperature. However, if the steel is heated in the range of 500°C–750°C, the carbides precipitate at the grain boundaries, thereby making the steel susceptible to intergranular corrosion as it depletes chromium from the adjacent grains. Thus, depending upon the composition, the stainless steels behave differently to heat treatment and accordingly are classified into three general groups.

12.1.1 FERRITIC STAINLESS STEELS

These are iron–chromium alloys having chromium approximately from 12% to 27%. These may contain small addition of manganese, silicon, nickel, molybdenum,

---

TABLE 12.1
Various Series of Stainless Steels

Series Designation	Groups
2xx	Chromium–nickel–manganese, nonhardenable, austenitic, nonmagnetic
3xx	Chromium–nickel, nonhardenable, austenitic, nonmagnetic
4xx	Chromium, hardenable, martensitic, magnetic
4xx	Chromium, nonhardenable, ferritic, magnetic
5xx	Chromium, low chromium, heat resisting

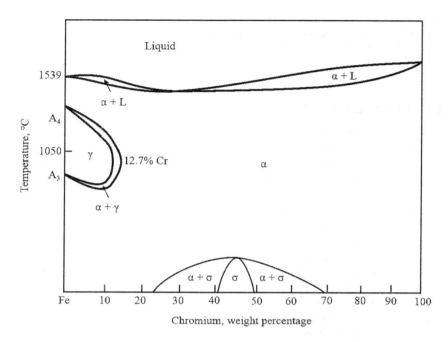

FIGURE 12.1 Iron–chromium phase diagram.

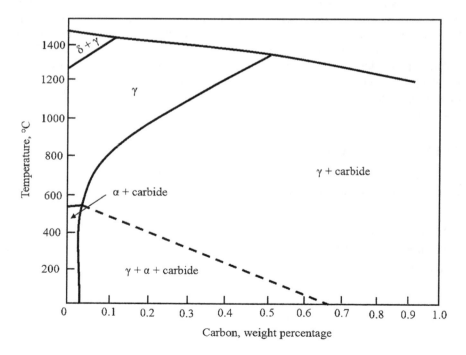

FIGURE 12.2 Influence of carbon on the phase diagram of 18/8 stainless steel.

Special Steels
171

and titanium. The carbon content is low (0.08%–0.2%), which helps to increase toughness and reduce sensitization. Annealing is the only heat treatment practice applied to such types of steels. In the annealed state, the strength of these ferritic stainless steels increases by ~50% higher than that of the carbon steels. These steels are extensively used for deep drawing applications such as vessels, kitchen sinks, and utensils. However, this variety of stainless steel suffers from brittleness upon slow cooling in the temperature range of 400°C–500°C. The reason behind the decrease in ductility as well as impact properties is probably due to the precipitation of very fine coherent chromium-rich, alpha prime particles arising from the miscibility gap in the Fe–Cr system. The effect is significant at higher chromium content.

12.1.2 MARTENSITIC STAINLESS STEELS

These are heat-treatable stainless steels and typically contain 12%–18% chromium and 0.10%–1.20% carbon. They have excellent corrosion resistance and toughness and can be easily hot-worked. These steels are hardened and tempered to obtain the best corrosion resistance but are not as good as the ferritic stainless steel. Like plain carbon steels, the strength and hardness of these steels also depend on the carbon content. The only difference is that the high alloy content of the stainless steels causes the transformation to be slow, and hence, maximum hardness is produced by air cooling. Based on the carbon content, these steels are further classified into two categories: low-carbon martensitic stainless steels and high-carbon martensitic stainless steels. Low-carbon martensitic steels are applicable where good weldability, formability, and impact strength are required. These steels are heated to above 1050°C to obtain a fully austenitic structure, then quenched in oil or air, and finally tempered. Tempering is done at a higher temperature (>600°C) to obtain higher toughness, although there is a slight reduction in corrosion resistance due to the precipitation of some carbides. In contrast, high-carbon martensitic stainless steels possess high strength and hardness due to the presence of more carbon, but at the expense of toughness and weldability. Higher austenitizing temperature is required to dissolve the excess carbides formed, which ultimately results in poor impact properties.

12.1.3 AUSTENITIC STAINLESS STEELS

Austenitic stainless steels have 16%–25% chromium and austenite-stabilizing elements such as nickel, manganese, or nitrogen. Typical examples include chromium–nickel (3xx type) and chromium–nickel–manganese stainless steels (2xx type). These steels find a wide range of applications due to excellent corrosion resistance than the ferritic and martensitic stainless steels. Type 304 steels, having a carbon content of 0.08% maximum, have good weldability and are used for chemical and food processing equipment due to the restriction in carbide precipitation during welding. Later, type 304L stainless steel is developed to avoid further carbide precipitation during welding, which contains only 0.03% carbon. Type 316 stainless steels show better corrosion resistance property than 302 or

172 Phase Transformations and Heat Treatments of Steels

304 steels because they contain 2%–3% molybdenum, and hence, they are used for pulp-handling purposes and photographic and food equipment. However, owing to the presence of high nickel content, type 3xx stainless steels are expensive. Alternatively, chromium–nickel–manganese stainless steels (type 201 and 202) have been developed with the substitution of manganese for nickel. The presence of manganese reduces the rate of work hardening.

12.2 HADFIELD MANGANESE STEELS

Hadfield manganese steel contains manganese above 10%. This steel is heated to around 1100°C, and then it is water quenched to obtain high strength and ductility, and excellent resistance to wear. These steels find applications in power shovel buckets and teeth, gears, spline shafts, axles, riffle barrels, grinding, and railway track work. If this steel is slow-cooled from 950°C, the structure consists of large brittle carbides neighboring the austenite grains, thereby decreasing the strength and ductility. Therefore, such type of steel is always heated to above 1000°C and then quenched to have much higher ductility and strength as compared to the annealed condition. The steel is ordinarily reheated to below 260°C to reduce the quenching stresses.

12.3 HIGH-STRENGTH LOW-ALLOY OR MICROALLOYED STEELS

High-strength low-alloy (HSLA) steels are prepared to have high fuel efficiency by offering automobile components with high strength-to-weight ratios. These are mild steels with carbon varying from 0.03% to 0.15%, manganese around 1.5%, and small amounts of other alloying elements. Such types of steels are subjected to controlled rolling to obtain ultrafine grains of size below 5 μm to attain a yield strength of about 550 MPa. The presence of low carbon with ultrafine ferrite grains is responsible for high strength and good weldability.

12.4 TRANSFORMATION-INDUCED PLASTICITY STEELS

These steels are produced by strain-induced transformation of austenite to martensite. Martensite forms in the metastable austenitic steels by the application of plastic deformation at room temperature. The steel composition is so chosen that its M_d temperature is above room temperature, such as 0.3% C, 2% Mn, 2% Si, 9% Cr, 8.5% Ni, and 4% Mo. The steel is then heated to 1200°C and quenched. It is heavily cold-worked above the M_d temperature (450°C) to deform the austenite and allow the precipitation of fine carbides to happen. This thermomechanical treatment with fine precipitation increases the M_d temperature. The steel is finally cold-rolled to below M_d temperature, i.e., at room temperature. It ultimately raises the M_s temperature of the steel, and hence, some austenite transforms to martensite. Depending upon the heat treatment steps, yield strengths from 700 to 2200 MPa with 90% to 20% ductility can be obtained. These steels are used where extremely high mechanical properties are desired, like in flat or wire forms.

Special Steels 173

12.5 MARAGING STEELS

Maraging steels are highly alloyed and low-carbon iron–nickel martensites that offer an excellent combination of strength, ductility, and toughness. These are mostly used for aerospace applications. Because of low carbon content, the martensite formed is soft with about 30 HRc. Hence, age hardening is done to strengthen the steel further. The composition of maraging steels is so chosen that lath martensite forms upon cooling the steel down to the room temperature. It is because lath martensite has several dislocations that are uniformly distributed and hence support age hardening by providing several preferred nucleation sites for the intermetallic compounds to precipitate. Besides, lath martensites provide favorite diffusion paths for substitutional solute atoms during aging. The M_s temperature of such types of steels is kept high around 200°C–300°C to avoid the formation of retained austenite. In some cases, nitriding of maraging steels is done to induce hardness of about 60–70 HRc up to a depth of 0.15 mm.

12.6 DUAL-PHASE STEELS

HSLA steels offer high strength-to-weight ratios, but their formability is poor. Hence, they are not suitable for deep cold pressing. HSLA exhibits a sharp yield point (stretcher strain), indicating nonuniform plastic deformation behavior. As a result, in cold-worked steel, surface finish is poor, and hence, these are not suitable for automobile applications.

Dual-phase steels overcome these problems. They have microstructures consisting of islands of martensite (10%–20%) or distribution of martensite in a ferrite matrix. It is a product of intercritical annealing, i.e., the steel is heated to a temperature of about 790°C in between A_1 and A_3 temperatures to obtain a microstructure consisting of ferrite and austenite, as shown in Figure 12.3. On quenching, austenite transforms into fine martensite. It is then tempered at low temperatures to improve the ductility. The formation of martensite depends on the alloy and carbon content of the austenite and the rate of cooling. Dual-phase steels offer several advantages such as low cost, excellent formability, a high strength of about 500–700 M/Nm², a low yield stress of about 300–350 MPa, and no yield point elongation.

12.7 TOOL STEELS

Tool steels are high-quality special steels used for forming and cutting purposes. They possess high hardness and wear resistance with economic tool life and dimensional stability. There are a variety of tool steels available ranging from cheap high-carbon steels to expensive high-speed steels, depending upon the applications. It is the alloying elements that improve the properties and thereby increase the cost of the tool steel. Various methods of classifying tool steels are summarized in Table 12.2. The tool steels are rated according to different properties such as toughness, red hardness, wear resistance, nondeforming properties, machinability, safety in hardening, and resistance to decarburization. Based on the application areas, tool steels with particular properties are selected.

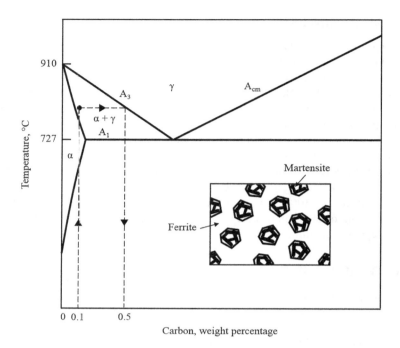

FIGURE 12.3 Schematic illustration of obtaining a dual-phase steel.

TABLE 12.2
Classification of Tool Steels

Criteria	Types
Quenching media	Water hardening
	Oil hardening
	Air hardening
Alloy content	Carbon tool
	Low-alloy tool
	Medium-alloy tool
Application	Hot work
	Shock resisting
	High speed
	Cold work

12.7.1 Water-Hardening Tool Steels

These are mostly plain carbon tool steels with carbon varying from 0.60% to 1.40%. They also contain some quantities of chromium and vanadium, which help to enhance hardenability and wear resistance. Depending upon the carbon content, these steels are further subdivided into three categories, as given in Table 12.3.

Special Steels

TABLE 12.3
Classification of Water-Hardening Tool Steels

Carbon Content	Properties	Applications
0.60%–0.75%	High toughness and wear resistance	Hammers, river sets, and concrete breakers
0.75%–0.95%	Good toughness as well as good wear resistance	Chisels, punches, dies, and shear blades
0.95%–1.40%	High wear resistance	Woodworking tools, drills, taps, reamers, and turning tools

These steels are generally water-quenched for obtaining high hardness and are given spheroidization annealing treatment to enhance the machinability. They have low red hardness and, hence, cannot be used as cutting conditions in cases where a considerable amount of heat is generated at the edge.

12.7.2 Shock-Resisting Tool Steels

Shock-resisting steels have high toughness and, as the name suggests, can withstand shock. They contain carbon between 0.45% and 0.65%. The primary alloying elements in these steels are silicon, chromium, and tungsten. Silicon strengthens and toughens the ferrite, while chromium rises the hardenability and wear resistance. Tungsten improves the red hardness property of these steels. Sometimes, molybdenum is added to increase the hardenability. The steels in this group are used in the manufacturing of forming tools, punches, chisels, pneumatic tools, and shear blades.

12.7.3 Cold-Work Tool Steels

This group of steels is the most essential group of tool steels, as it offers a wide range of applications. The key alloying elements added to this group of steels are chromium, vanadium, and tungsten. High toughness, abrasion and wear resistance, and impact resistance are some of the vital properties of cold-worked steels. These steels are subdivided further into three classes – oil-hardening steels, air-hardening steels, and high-carbon high-chromium steels.

Oil-hardening steels contain manganese, chromium, and tungsten. They enhance hardness and wear resistance. As the name suggests, these are hardened by oil quenching. They possess excellent nondeforming properties and are, therefore, unlikely to bend, distort, twist, or crack during the heating process. Sometimes, silicon is added to improve the machinability in the annealed state and increase the resistance to decarburization of the steels. They are used for taps, form tools, and expansion reamers.

Air-hardening steels are hardened by air cooling. They mostly contain about 1% carbon, 3% manganese, 5% chromium, and 1% molybdenum. Among the alloying elements, manganese and molybdenum impart air-hardening properties and induce high hardenability. They have excellent nondeforming properties, wear resistance,

176 Phase Transformations and Heat Treatments of Steels

toughness, red hardness, and resistance to decarburization. These are mostly used for blanking, forging, and trimming purposes.

High-carbon high-chromium steels contain carbon up to 2.25% and 12%–14% chromium. Besides, they may contain other alloying elements such as molybdenum, vanadium, and cobalt. They offer superb wear resistance, nondeforming properties, excellent abrasion resistance, and minimum dimensional change in hardening. This group of steels are used for blanking and piercing dies, drawing dies for wires, bars, shear blades, and cold-forming rolls.

12.7.4 Hot-Work Tool Steels

This group of steels is mostly applicable for high-temperature metal-forming processes such as hot forging, hot stamping, hot extrusion, die casting, and plastic molding. The operating temperature varies from 200°C to 800°C. They have good red hardness due to the alloying elements such as chromium, molybdenum, and tungsten. The red hardness property becomes significant when the sum of these alloying elements is at least 5%. Besides, they possess high wear resistance, toughness, erosion resistance, and resistance to softening at high temperatures. Depending upon the alloying elements, hot-worked tool steels are subdivided into three categories – hot-work chromium-base, hot-work tungsten-base, and hot-work molybdenum-base.

Hot-work chromium-base steels usually contain a minimum of 3.25% chromium and minor amounts of other alloying elements such as tungsten, vanadium, and molybdenum. Because of these alloying elements, they have high red hardness and high hardenability. These are mostly used for extrusion dies, die-casting dies, and hot shears. A specific hot-work tool steel, H11 tool steel, finds applications in highly stressed structural parts, predominantly for supersonic aircraft, as it can resist softening upon continuous exposure to temperatures up to 500°C.

Hot-work tungsten-base steels have at least 9% tungsten and 2%–12% chromium. With higher alloying elements content, these steels have high resistance to high-temperature softening as compared to hot-work chromium-base steels. These are used for mandrels, punches, and extrusion dies for brass, nickel alloys, and steels.

Hot-work molybdenum-base steels contain 8% molybdenum, 4% chromium with small amounts of tungsten, and vanadium. These steels possess almost similar characteristics as those of hot-work tungsten-base steels but have higher toughness because of low carbon content. They are poor in resistance to decarburization, and therefore, proper care needs to be taken during heat treatment.

12.7.5 High-Speed Tool Steels

High-speed steels contain large amounts of tungsten or molybdenum along with some chromium, vanadium, and cobalt. They consist of carbon varying from 0.70% to 1.5%. They have good nondeforming properties, wear resistance, machinability, and resistance to decarburization. These are mostly used for cutting tools, burnishing tools, preparing extrusion dies, and blanking punches and dies. Based on the alloying elements content, these steels are subdivided into two groups – tungsten-base high-speed steels and molybdenum-base high-speed steels. Among the tungsten-base

Special Steels 177

high-speed steels, 18-4-1 (T1) is the most commonly used high-speed steel, which contains 18% tungsten, 4% chromium, and 1% vanadium. Since tungsten is expensive, it is replaced by molybdenum, which gives rise to molybdenum-base high-speed steels. Sometimes, cobalt is added when high red hardness is required. For example, T15 steel, which contains vanadium and cobalt, provides dominance in both red hardness and abrasion resistance. However, the addition of cobalt makes the steel brittle, and hence, the high-cobalt steels must be protected against excessive shock or vibration during service.

12.8 ELECTRIC GRADE STEELS

Electric grade steels may contain up to 6.5% silicon, but commercial grades usually have silicon content up to 3.2% because higher concentrations may lead to brittleness during cold rolling. Also, other alloying elements such as manganese and aluminum up to 0.5% are added. While silicon helps in lowering the core loss, carbon causes magnetic aging by precipitating carbides. Thus, carbon content is kept very low to around 0.005%. Electrical steels are generally manufactured in cold-rolled strips, which are used as cores of transformers, stator, and rotor of electric motors. There are two varieties of electrical steels – cold-rolled grain-oriented steel (CRGO) and cold-rolled non–grain-oriented steel (CRNGO). While CRNGO steels have similar magnetic properties in all directions, CRGO steels are treated in such a way that the optimal features are developed only in the rolling direction. CRGO steels are mostly used in the cores of power and distribution transformers because of their ability to increase the magnetic flux density in the coil rolling direction. On the other hand, CRNGO electrical steels find applications where magnetic flux density is not constant, such as in electric motors and generators with moving parts.

FURTHER READING

Avner, S. H. *Introduction to Physical Metallurgy.* (Tata McGraw-Hill Education, New York, 1997).
Singh, V. *Heat Treatment of Metals.* (Standard Publishers Distributors, New Delhi, 2006).
Rajan, T. V., Sharma, C. P., & Sharma, A. *Heat Treatment: Principles and Techniques.* (PHI Learning Pvt. Ltd., New Delhi, 2011).

13 Some *In Situ* Postweld Heat Treatment Practices

Very often, it is required to join steel components in order to prepare a structure. Automobile industries are the major players in this regard. Welding of sheet metal is one of the critical concerns for automobile industries, as it targets for improved crashworthiness and fuel economy. In fact, welding is unavoidable in almost every structural application, may it be mobile structures such as automobile and railways or immobile structures such as roofs, bridges, and so on. However, welding causes structural distortions to the components and usually the weld zone, which is known as heat-affected zone (HAZ), and more specifically, fusion zone (FZ) exhibits poor mechanical performance than that of the unwelded structure. Usually, this FZ failure is a common phenomenon during a collision, which reduces the crashworthiness of the automobile and thus should be taken care of in order to have a reliable and durable long-term application. Welding is not only a requirement while fabricating a new structure but also essential during maintenance or replacement of a component in an assembly. In order to restore the properties of the metal at FZ by modifying the microstructure and relieving the internal stresses, certain postweld heat treatment (PWHT) is essential. In case of certain pressure vessels and pipes, PWHT is mandatory to ensure safe operation of the components in both metallurgical and mechanical aspects.

13.1 NECESSITY

During a fusion welding operation, a part of the component is melted and resolidified. This process takes place only at a very small localized region of the component. The high heating rate associated with the welding operation followed by high cooling rate brings several changes in the microstructure in the HAZ and thus exhibits different characteristics than that of the base metal. This evolved temperature gradient also leads to generation of residual stresses, which is also required to be released in order to preserve the performance of the welded joints. These residual stresses further become very critical in case of thick cross-sectioned components. In certain cases, where the cooling rate surpasses a critical value, martensitic transformation also may take place reducing the ductility of the joint drastically. In certain sensitive microstructures, hydrogen-induced cracking (HIC) also may be resulted with substantial amount of hydrogen and applied stress. Hence, it is a usual practice to perform PWHT in critical applications, where unprecedented failure is not tolerable.

13.2 CONVENTIONAL POSTWELD HEAT TREATMENT PROCESS

Typically, after welding, the component is heated to an elevated temperature followed by some period of holding and then controlled cooling. Two types of PWHT are normally practiced: one in which there is a possibility of HIC, known as *postheating*, and the other in which a high magnitude of residual stress exists, known as *stress relieving*.

13.2.1 POSTHEATING

Combination of a sensitive microstructure along with a substantial amount of hydrogen and structural load gives rise to a possibility of hydrogen-induced cracking. In general, this hydrogen embrittlement usually happens close to the ambient temperature. Hence, suitable heat treatment must be done immediately after welding before the structure reaches the room temperature. From an intermediate temperature (just after welding), the steel should be heated to the postheating temperature (~200°C–300°C) followed by a suitable residence time and then cooling.

13.2.2 STRESS RELIEVING

This heat treatment is adopted when the steel structure experiences a high magnitude of residual stress as a consequence of fabrication process. In dome of the fabrication/welding operations, the residual stresses may even go rise to the yield strength of the material. Hence, for suitable in-service performance, these residual stresses must be released. A slow heating is often required to a temperature below the lower critical temperature of the steel followed by some holding time and then slow cooling. Most of the time, the holding temperature is 600°C–700°C, and holding time is usually 1 h for 15 mm thickness. Stress relieving benefits the structure by several possible ways, i.e., by improving the dimensional stability and reducing the possibility of stress corrosion cracking.

13.3 *IN SITU* POSTWELD HEAT TREATMENT OF TRANSFORMATION-INDUCED PLASTICITY STEEL

As discussed in the earlier chapter, transformation-induced plasticity (TRIP) steel is a special steel having a combination of ferrite, martensite, and retained austenite. The TRIP effect is due to transformation of austenite to martensite upon loading. Hence, under tensile kind of loading, the tendency of necking is minimized with improved rate of strain hardening. This also enhances the energy absorption capacity of the material. While all these aforementioned characteristics are desirable for automobile applications, weldability of these steels overshadows its application in such area. Interfacial failure through crack propagation across the FZ reduces its load bearing energy absorption ability, while loaded in a cross tension mode. The reduced weldability of such TRIP steel may be attributed to the brittle martensite formed in the FZ. Very often, resistance spot welding is carried out for such steels in various automobile applications. In situ PWHT thus would be an appropriate method in order to restore the physical and mechanical properties of the weldment. The martensite formed at the FZ and HAZ is required to be tempered in order to reduce the brittleness. This is conventionally achieved by appropriate hardening and tempering operations. The weldment should come to a temperature below the M_f temperature before start of the tempering process. In certain cases, reduction of hardness of the HAZ is not the only requirement. In addition, if there is incoherency among the dendritic structure after solidification due to segregation of S and P, then the edges of the weldment should be remelted to modify the unwanted dendritic structure. Here, let us discuss a suitable technique for in situ PWHT of TRIP steels after resistance spot welding.[1]

A welding operation was carried out using resistance spot welding using a PLC-controlled machine. This process produced a weldment with a conventional recommended FZ size of $4\sqrt{t}$. This process was followed by application of a second pulse current (3.5, 5.5, and 7.5 kA amplitude) for PWHT purpose.

Figure 13.1a displays a conventional welding microstructure comprising FZ, HAZ, and base metal (BM). The microstructure of the metal in FZ was found to be martensite as shown in Figure 13.1b. This martensite is quite hard with hardness 425 HV, much higher than that of the base metal. Mode I loading during CT testing further indicates failure of the joint by propagation of the crack into the FZ, resulting in interfacial failure as evident from Figure 13.1c. Semicleavage fracture mode can be noticed from Figure 13.1d indicating the presence of hard brittle phase in the FZ. Furthermore, the carbon equivalent of the steel plays an important role in driving the crack to the FZ.

Effect of PWHT with different current is shown in Figure 13.2. All the PWHT operations have increased both the peak load and failure energy substantially in comparison to the as-welded structure with maximum of 60% and 50%, respectively.

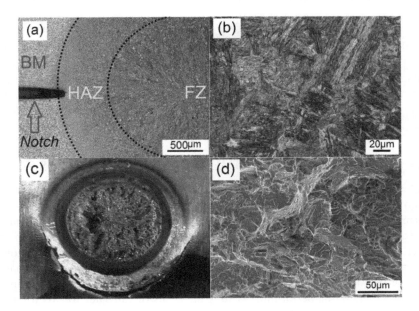

FIGURE 13.1 (a) Weldment comprising FZ, HAZ, and BM, (b) martensite in the FZ, (c) interfacial failure, and (d) fractography of the fractured surface.[1] FZ, fusion zone; HAZ, heat-affected zone.

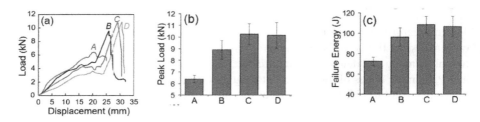

FIGURE 13.2 Effect of PWHT on CT test, (a) load vs. displacement, (b) peak load, and (c) failure energy.[1] (A: as-welded, B: PWHT with 3.5 kA, C: PWHT with 5.5 kA; D: PWHT with 7.5 kA). PWHT, postweld heat treatment.

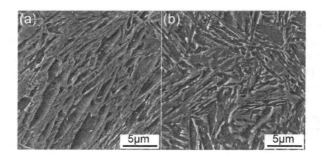

FIGURE 13.3 Microstructure of FZ in its (a) as-welded and (b) tempered conditions.[1] FZ, fusion zone.

Visual inspection of the fractured specimens indicates pullout failure to be the dominating failure mode rather than interfacial failure. This may be attributed to transformation of the hard and brittle martensite at the HAZ to relatively soft and ductile tempered martensite as shown in Figure 13.3. Lath martensite present in the FZ after welding (Figure 13.3a) has been transformed to tempered martensite (Figure 13.3b) after performing the PWHT process.

13.4 POSTWELD HEAT TREATMENT OF DUPLEX STAINLESS STEEL

Duplex stainless steels are optimal combination of mechanical properties and corrosion resistance and very often the preferred structural material in various load bearing applications serving at harsh and corrosive environments, such as oil and gas industries, marine, chemical container, waste drainage pipes, and so on. The microstructural constituents, i.e., ferrite and austenite render this special combination of properties in these duplex stainless steels. For achieving this optimal combination of properties, ideally a 50% ferrite–50% austenite is required to be present in the microstructure. While performing welding on such structures due to any in-service or assembly or maintenance requirement, this ratio is altered due to associated phase transformation. Due to presence of substantial high volume fraction of the alloying elements, these steels tend to form several undesirable phases such as chi phase,

In Situ Postweld Heat Treatment Practices

carbide phase, sigma phase, secondary austenite, and so on in addition to ferritization, which may negatively affect the properties and performance of the weldment. Hence, suitable PWHT operation is essential for long-term reliable applications. Here, the effects of short PWHT on the properties of a tungsten inert gas (TIG)–welded duplex stainless steel are discussed.[2] Cold-rolled duplex stainless steel sheets of 1.5 mm thickness were taken and welded by single TIG welding method without use of any filler material. This is followed by heat treatment of the joints in a muffle furnace at 1000°C–1100°C (in the range of temperature where precipitation does not take place) in normal atmosphere with a holding time of 1.5 min followed by water quenching.

The base metal (Figure 13.4a) comprising ferrite (bright region) and austenite (dark region) is a signature of duplex stainless steel having austenite embedded in a continuous ferritic matrix. Columnar ferrite and equiaxed ferrite grain are visible in the FZ at near the fusion line and center, respectively, as shown in Figure 13.4b and c. Widmanstätten-type austenite (WA), intragranular austenite (IGA), and grain boundary austenite (GBA) are embedded in the ferrite grains as shown in the figures. The HAZ contains a gradient of microstructural changes due to temperature gradient in the region from melting point of the metal near the FZ to ambient temperature at the base metal along with differential heating and cooling rates.

The microstructure of the FZ after the PWHT was also obtained in order to verify the possible changes in the microstructural features as shown in Figure 13.5. Widmanstätten austenite and allotriomorphs austenite at the grain boundaries tend

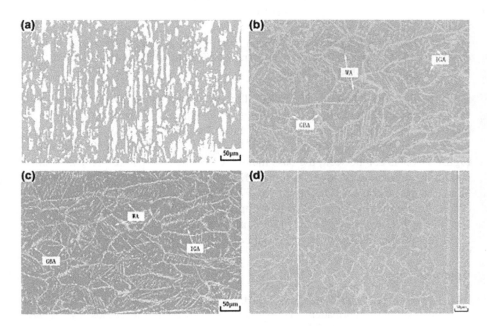

FIGURE 13.4 Microstructure of duplex stainless steel, (a) base metal, (b) FZ near the fusion line, (c) center of FZ, and (d) HAZ (in between two white vertical lines).[2] FZ, fusion zone; HAZ, heat-affected zone.

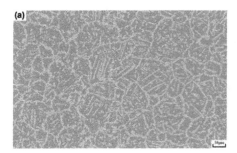 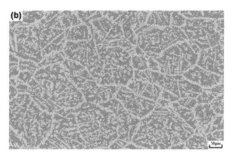

FIGURE 13.5 Microstructure of the FZ after PWHT at (a) 1020°C and (b) 1080°C for 15 min.[2] FZ, fusion zone; PWHT, postweld heat treatment.

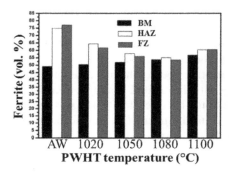

FIGURE 13.6 Ferrite content in base metal (BM), HAZ, and FZ in the as-welded (AW) condition and after PWHT at different temperatures.[2] FZ, fusion zone; HAZ, heat-affected zone; PWHT, postweld heat treatment.

to increase as the PWHT is increased. This increment indeed is associated with spheroidization of the intragranular austenite. Overall, all these facts point out to have a higher austenite content on the FZ due to the PWHT process.

The ferrite content at different zones of the welded metal after PWHT is shown in Figure 13.6. As can be seen, the base metal has almost 50% ferrite in the as-welded condition and increases slowly as the PWHT temperature is increased. On the other hand, the ferrite content is highest in the FZ followed by HAZ in the as-welded condition. This gradually decreases as the temperature of PWHT is elevated till 1080°C, and all the zones more or less contain equivalent ferrite content. However, beyond that temperature, the ferrite content further tends to increase.

REFERENCES

1. Sajjadi-Nikoo, S., Pouranvari, M., Abedi, A., & Ghaderi, A. A. In situ postweld heat treatment of transformation induced plasticity steel resistance spot welds. *Sci. Technol. Weld. Joining* **23**, 71–78 (2018).
2. Zhang, Z., et al. Effect of post-weld heat treatment on microstructure evolution and pitting corrosion behavior of UNS S31803 duplex stainless steel welds. *Corros. Sci.* **62**, 42–50 (2012).

14 Heat Treatment of Cast Iron

14.1 INTRODUCTION

Cast irons are mostly alloys of iron and carbon that contain carbon between 2.1% and 6.67%. At the eutectic temperature, such high levels of carbon are essential to saturate austenite so that carbon precipitates in the form of carbides and graphite in the structure of the alloy. Being brittle, they cannot be forged, rolled, drawn, or worked at room temperature. However, they can be quickly melted and "cast" into desired shapes, and hence, they are commonly known as cast irons. Melting units for producing liquid melt for casting may be an electric arc, cupola, reverberatory, or induction furnace.

Typical cast irons are cheap and, therefore, are of importance, as the properties of cast iron can be varied over a broader range by proper alloying, reasonable foundry control, and suitable heat treatment practice. Although they are brittle and have less strength compared to steel, they have other useful features.

14.2 TYPES OF CAST IRON

The most appropriate method of classification of cast iron is dependent on the metallographic structures. Carbon content, the alloy and impurity content, the rate of cooling during and after solidification, and the heat treatment after casting are four variables that need to be considered for differentiating the cast irons from each other. These variables control the physical form and condition in which carbon is present in the alloy. Carbon can present in a combined state as iron carbide or cementite or occur as free carbon in graphite. Physical and mechanical properties of cast iron are greatly influenced by the shape and distribution of the carbon in the structure of cast irons. The classification of cast irons which is best done based on their microstructure is as follows:

- White cast iron: Carbon exists in the combined form as cementite.
- Gray cast iron: Carbon exists in the uncombined form, such as graphite flakes.
- Malleable cast iron: It is a product of heat treatment of white cast iron, and carbon is present in uncombined form as irregular round particles, also called as temper carbon.
- Spheroidal graphite (S.G.) iron or nodular iron: Alloy addition leads to the formation of spheroids of uncombined carbon during cooling. It is unlike malleable cast iron, where temper carbon is formed not only during cooling but also during heat treatment after casting.

185

14.2.1 WHITE CAST IRON

According to the best method of classification of cast iron via metallographic structure, white cast iron is one in which all the carbon is present in a combined state, also known as cementite. They have more than 2.11% carbon and consist of hard carbide phases surrounded by a hardenable metal matrix. Both constituents grow from the melt and differ in crystallographic structure, and therefore, the hardness depends on alloying and heat treatment. White cast irons have been the doers of wear protection in the mining and cement industry, as well as in road building where abrasion by mineral grains prevails.

The changes occurring in the microstructure during solidification and subsequent cooling can be corroborated from the iron–iron carbide phase diagram, as discussed in Chapter 6. A hypoeutectic white cast iron containing about 2.5% carbon when cooled from the molten state (point 1 in Figure 14.1) under equilibrium conditions begins solidification at the liquidus line with the formation of dendrites of austenite (point 2). Solid dendrites of proeutectic austenite continue to form by losing carbon to the liquid until eutectic temperature (1147°C). At this point (point 3), the alloy consists of proeutectic austenite having 2.11% carbon and liquid having 4.3% carbon. The liquid undergoes eutectic reaction forming ledeburite, which is

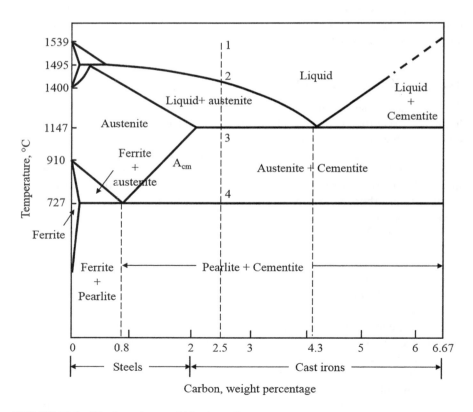

FIGURE 14.1 The iron–iron carbide phase diagram.

Heat Treatment of Cast Iron

a eutectic mixture of austenite and cementite. Usually, this newly formed eutectic austenite gets added to the primary austenite leaving behind massive layers of cementite. The A_{cm} line of the iron–iron carbide phase diagram indicates that the solid solubility of carbon in austenite decreases with temperature (1147°C–727°C), and this results in the deposition of proeutectoid cementite (from both the primary and eutectic austenite) onto the already-existing eutectic cementite. The dendrites of primary austenite break due to the secondary (proeutectoid) cementite precipitating out from them. At the eutectoid temperature (727°C), all the austenites having 0.77% carbon undergo eutectoid reaction forming pearlite, i.e., ferrite and cementite. The structure essentially remains unchanged after cooling from this temperature to room temperature. The stages of formation of white cast iron are schematically illustrated in Figure 14.2.

The characteristic microstructure of white cast iron consists of black dendrites of transformed austenite (pearlite) in a white interdendritic network of cementite. Cementite is brittle and has a high hardness (~ 1000 VPN). Large amounts of cementite make the white cast irons hard and wear resistant, but because of extreme

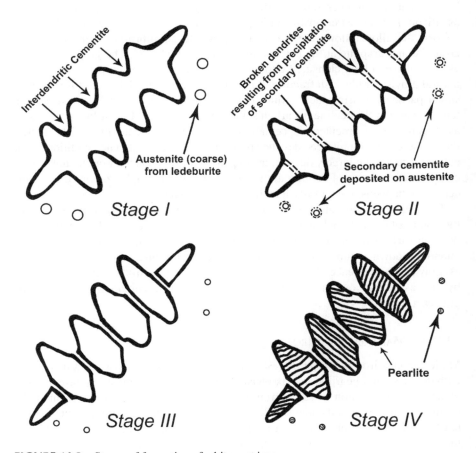

FIGURE 14.2 Stages of formation of white cast iron.

188 Phase Transformations and Heat Treatments of Steels

brittleness, these are difficult to machine. Therefore, they are only used where resistance to wear is mostly needed, and ductility is not a necessary material attribute, like liners used in cement mixers, extrusion nozzles, ball mills, and specific types of drawing dies.

14.2.2 GRAY CAST IRON

Gray cast irons have flakes of graphite embedded in a matrix of steel. They show a gray-blackish-colored fracture. The size and character of the steel matrix play an essential role in the strength of gray cast iron. Gray cast irons are hypoeutectic cast irons having a total carbon content between 2.4% and 3.8%. Silicon content is between 1.2% and 3.5%. By adequately controlling the composition and cooling rates, cementite tends to separate into graphite and austenite or ferrite.

Initially, these alloys solidify to form primary austenite. The first appearance of combined carbon in the form of cementite occurs from the eutectic reaction at 1143°C. However, the graphitization process is aided by various factors like high carbon content, high temperature, and the proper amount of graphitizing elements, such as silicon. It has been experimentally proved that with proper control of the above factors, the alloy follows the stable iron–graphite equilibrium diagram, forming austenite and graphite at the eutectic temperature of 1143°C. The graphite appears as several irregulars, elongated and rounded plates that provide gray cast iron its natural grayish or blackish fracture. It may be noted that the flakes are three-dimensional particles. During the continued cooling process, additional carbon precipitates because of the decline in carbon solubility in austenite. The carbon precipitates as graphite or as proeutectoid cementite, which quickly graphitizes. The matrix that embeds the graphite decides the strength of the gray cast. The condition of the eutectoid cementite largely determines the nature of this matrix. For example, if the composition and rate of cooling are so adjusted that the graphitization of eutectoid cementite happens, then a fully ferritic matrix is produced.

In contrast, if the eutectoid cementite graphitization is avoided by controlling the composition and cooling rate, the matrix is entirely pearlitic. The matrix composition varies from pearlite, through the mixtures of pearlite and ferrite in various quantities, down to pure ferrite. The graphite–ferrite mixture is the weakest and softest gray cast iron. Hence, the hardness and strength of this mixture can be increased by increasing the combined carbon until it reaches a maximum with the pearlitic gray cast iron.

14.2.3 MALLEABLE CAST IRON

Malleable cast iron is a type of cast iron whose structure consists of a metastable carbide phase in a pearlitic matrix phase. This metastable cementite phase tends to undergo decomposition to stable iron and carbon phases. It does not occur under normal conditions, and the metastable carbide phase persists indefinitely in its original form. Cementite phase takes a long time to decompose into stable phases, and the reaction $Fe_3C \rightleftharpoons 3Fe + C$ is favored by several factors, including high temperatures, high carbon content, presence of impurities, and elements that support Fe_3C

Heat Treatment of Cast Iron

decomposition. The stable iron–carbide (graphite) system put on top of the metastable systems can be seen in Figure 14.3. The ideal composition that is suitable for the conversion of white iron to the malleable cast iron is given in Table 14.1.

Production of malleable cast iron requires the conversion of the carbon present in the white cast iron into two separate phases of ferrite and irregular nodules of tempered carbon (graphite). This conversion takes place in two stages of annealing. In the first stage, annealing is done where the white cast iron is heated to a temperature range of 900°C–950°C. During heating, the pearlite phase is converted to the austenite phase. The austenite dissolves an additional cementite phase as its temperature reaches the austenitization temperature. It is shown in Figure 14.2 that the austenite formed in the metastable iron–iron carbide system tends to dissolve more carbon than the austenite of the stable iron–graphite system. As a result, a driving force arises, which results in the precipitation of the carbon as a free elemental graphite. This precipitation leads to the growth of the graphite nucleus, which grows at a uniform rate in all directions. These graphite precipitates appear as irregular nodular shapes or spheroids, commonly known as temper carbon. For the precipitation, graphite particles need to nucleate and then further grow, which depends on many factors, including the presence of any defects, foreign particles, and nucleating surfaces.

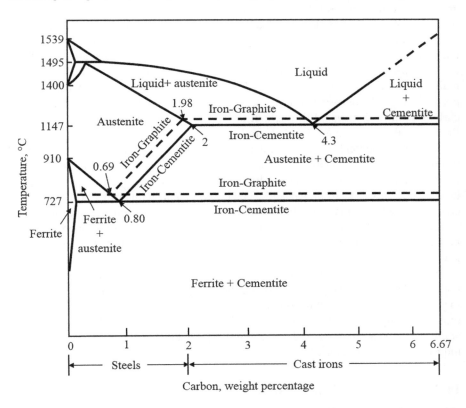

FIGURE 14.3 The iron–graphite system (dotted lines) placed over the iron–iron carbide phase diagram.

Phase Transformations and Heat Treatments of Steels

TABLE 14.1
Typical Chemical Composition of Malleable Cast Iron

Element	Composition (%)
Carbon	2.16–2.90
Silicon	0.90–1.90
Manganese	0.15–1.25
Sulfur	0.02–0.20
Phosphorus	0.02–0.15

Some elements like silicon tend to promote this reaction, but it should not exceed beyond a permissible limit; otherwise, the iron may not solidify as white cast iron, which is undesirable. These nuclei are obtained through proper annealing cycles and methods. Several parameters, such as temperature, composition, and nucleation tendency determine the rate of annealing. Increasing temperature has both positive and negative impacts. While increasing the temperature increases the decomposition tendency of the carbide phase leading to increased graphite precipitation, it has specific limits, above which distortions occur in the annealed product. Hence, the temperature needs to be chosen to account for these considerations and, therefore, are controlled between 900°C and 950°C. In this temperature range, the cast iron is maintained for an extended period until most of the primary carbide phase decomposes. The final microstructure at the end of the first stage consists of tempered carbon nodules distributed throughout the matrix of saturated austenite. After the first stage, the cast iron is cooled at a rather rapid rate.

During the second stage, the castings are cooled at a slow rate of 5–10°C/h through the eutectoid region. The carbon dissolved in the austenite phase converts to the graphite phase during the cooling process. These graphite phases get deposited on the tempered carbon particles. The remaining austenite transforms to ferrite. Once the graphitization is finished, no microstructural change takes place on further cooling to room temperature. The final microstructure consists of tempered carbon nodules in the ferritic matrix. This type of iron is known as standard or ferritic malleable iron.

When the carbon nodules are compact, discontinuity in the ferritic matrix does not take place, which helps in improving the strength and the ductility as compared to gray cast iron. The graphite nodules serve as lubrication purposes for use in cutting tools, resulting in good machinability of the tools. The ferritic malleable irons find applications in automotive and agricultural equipment, railing casting on bridges, industrial casters, and pipe fittings.

14.2.4 Spheroidal Graphite Iron

Spheroidal Graphite (S.G.) iron, also called nodular cast iron or ductile iron, contains spheroids of the graphite phase embedded in the steel matrix. The matrix can be

Heat Treatment of Cast Iron

ferritic or pearlitic, which governs the final properties of the casting. The general composition of S.G. iron is given in Table 14.2. The microstructure of S.G. iron, as shown in Figure 14.4, shows that the spheroids are more smooth-edged than the irregular aggregates of temper carbon of the malleable iron.

S.G. iron is produced by controlled treatment of the cast iron using spheroidizing elements like magnesium or cerium, which helps in nodule formation of the graphite phase. Based on the alloying proportion, there are different types of ductile irons. Ferritic irons are those who have a matrix with a maximum of 10% pearlite. This type of iron has maximum ductility, toughness, and machinability. The matrix can be changed to martensite by quenching in oil or water from 850°C to 950°C. The quenched structures are generally tempered to obtain the required strength and hardness.

On the other hand, austenitic ductile irons contain alloying elements responsible for maintaining the austenitic structure. These irons are useful where high corrosion resistance and excellent creep properties at high temperatures are required. The properties of different types of S.G. iron are given in Table 14.3.

TABLE 14.2
Typical Composition of Elements in Spheroidal Graphite Cast Iron

Element	Composition (%)
Carbon	3.2–4.1
Silicon	1.0–2.8
Phosphorus	<0.10
Sulfur	<0.03
Manganese	0.3–0.8
Magnesium	0.04–0.06

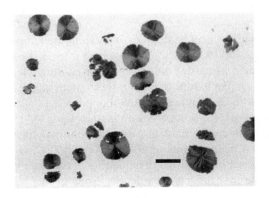

FIGURE 14.4 Microstructure of S.G. cast iron.[1] S.G., spheroidal graphite.

192   Phase Transformations and Heat Treatments of Steels

TABLE 14.3

Different Properties of Various Types of Spheroidal Graphite Iron

Type	Tensile Strength (MPa)	Hardness (BHN)	Elongation (%)	Heat Treatment
Ferrite	350	130	22	Annealed
Ferrite	420	150	12	Annealed
Ferrite + pearlite	500	275	7	Annealed
Tempered Martensite	800	320	2	Quenched and tempered
Austenitic	400	200	40	Cast

14.3 HEAT TREATMENT OF GRAY CAST IRON

Among the heat treatment cycles, stress relieving is probably the most frequently applied heat treatment to gray cast irons. As cooling proceeds at different rates through various sections of the casting, it leads to the development of residual stresses in the as-cast gray iron. These residual stresses reduce strength, cause distortions, and in extreme cases, may also lead to cracking. Hence, stress-relieving treatment is done to remove unwanted stresses. Stress-relieving temperature is generally well below the pearlitic to austenitic transformation range. These stresses are wholly removed by soaking the alloy in the range of 550°C–650°C.

Annealing treatment of gray cast iron is carried out to graphitize carbides and to homogenize the castings. The alloy is softened, and its ductility and machinability increase. At machine-building plants, gray iron castings are most frequently annealed to improve their machinability. The annealing treatment involves soaking the castings for up to 10 h at 850°C–950°C. Annealing of gray cast irons leads to cementite decomposition. The intensity of graphitization increases with heating temperature. Up to 600°C temperature, the treatment has an insignificant effect on the structure of gray iron. Above 600°C, there is a marked increase in the rate at which iron carbide decomposes to graphite and ferrite, which reaches a maximum at 760°C for unalloyed and low-alloy gray cast irons. Small castings of simple shapes may be heated to a temperature of 1050°C–1150°C in a salt bath for a minimum holding time for the decomposition of cementite.

Normalizing treatment may be done by increasing the temperature to above the transformation range (900°C–950°C) and soaking it for a particular time depending upon the thickness of casting followed by air cooling. Normalizing enhances mechanical properties like hardness and tensile strength and restore as-cast properties that have been altered by other heat-treating processes, such as graphitizing or preheating and postheating associated with repair welding.

Like steel, gray iron can be hardened by cooling rapidly or quenching in water, oil, or salt bath from temperatures of about 800°C–850°C. Oil quenching is preferred over water quenching for through hardening as the latter may cause distortion and cracking. For flame or induction hardening, water quenching is utilized, as only the outer surface needs to be hardened. Quenched iron can be tempered in the range

Heat Treatment of Cast Iron

of 150°C–650°C to relieve stresses and improve toughness. Although tempering enhances strength and toughness, it decreases hardness. A minimum temperature of about 370°C is required to restore the impact strength to the as-cast level. Heat treatment is not customarily used commercially to increase the strength of gray iron castings. Specific alloying elements can improve the strength of the as-cast metal. Besides, a reduction in the amounts of silicon and total carbon increases the load-bearing capacity of the castings. Gray iron is generally quenched and then tempered to increase the wear and abrasion resistance by increasing the hardness. The treatment produces a structure consisting of graphite embedded in a hard martensitic matrix. The combination of hardness and excellent wear resistance makes the gray cast irons suitable for some applications such as sprockets, diesel cylinder liners, farm implement gears, and automotive camshafts.

14.4 HEAT TREATMENT OF MALLEABLE CAST IRON

We have established in the earlier section that the malleable irons of pearlite and ferrite type are obtained after subsequent malleable iron formation and annealing treatment of the white cast iron where the primary carbide phase undergoes decomposition reaction. The procedure takes a considerable amount of time and involves a rapid rate of cooling; hence, specimen with large dimensions and broad cross section may become challenging to achieve the malleable cast iron morphology. Malleable cast irons are used in applications that require good malleability of the tool along with appreciable machinability. The following heat treatments can be given for malleable cast iron.

Increasing the austenitization time and temperature increases the dissolved carbon content in the austenite. It leads to the homogenization of the austenitic phase. However, high temperatures also create distortion and stress generation, thereby leading to crack formation. The suitable austenitization temperature for pearlitic malleable iron is 885°C and that for ferritic malleable cast iron is 900°C. The hardening steps for pearlitic malleable iron are as follows:

- The as-cast specimen are quenched in the air first after the first stage of graphitization and carbide decomposition. Uneven distribution of carbon may still be present in the matrix.
- The castings are again heated for 1 h at a suitable austenitization temperature, which leads to a homogenous structure.
- Oil quenching is done in a hot oil bath (50°C–55°C) with agitation, resulting in the formation of martensite of about 600 BHN hardness.
- Based on the hardness and ductility requirements, the castings are tempered by heating to the temperature range of 600°C–700°C for 2 h. It leads to the homogenization of properties. Heat treatment cycles are extremely temperature dependent, and the higher the temperature, the higher the tendency of carbon precipitation.

The ferritic malleable cast iron can also be given similar heat treatment cycles to produce the same microstructures. The austenitization temperature is slightly higher than that of pearlitic malleable cast iron, which is in the temperature range of 900°C–930°C. The holding or soaking time is also higher. The resulting casting has a lower carbon content than that of the pearlitic malleable iron. As a result, the tempering temperature is also at a lower level.

Martempering is another heat treatment given to specimens that are susceptible to crack formation due to stresses generated during quenching. The casting is austenitized at 885°C for 1 h, soaked in a salt bath at the temperature of 200°C followed by air cooling. Tempering of the castings can be done to get the desired ductility and strength. Bainite phase can be achieved after austempering pearlitic malleable cast iron, which leads to increased tensile strength, yield strength, and ductility. Surface hardening methods, including flame induction, can be used for increasing the hardness of the specimen containing small dimensions and parts. The operation parameters control the depth of hardening and composition. Other methods can also harden the desired area of both ferritic and pearlitic malleable cast irons.

14.5 HEAT TREATMENT OF SPHEROIDAL GRAPHITE IRONS

S.G. cast iron consists of a graphite phase embedded in the ferrite or pearlite matrix, which is dependent on the composition and other various factors. The nodular shape of the graphite phase helps in avoiding any stress raising effect, and hence, this proves beneficial in improving the mechanical properties like ductility and tensile strength.

The as-cast iron develops internal stresses during the casting process, which can be reduced by heating the casting and holding it at a suitable temperature. The time of holding is dependent on the composition, complexity, final properties desired, and dimensions of the cast. Stress-relieving temperatures for various irons are given in Table 14.4.

Higher temperatures help in eliminating the stresses, but with higher temperatures, hardness and tensile strength values decrease. After holding the castings at the given temperature, they are furnace cooled and then cooled in the air to avoid reintroduction of residual stresses.

Annealing treatment is required when good machinability and ductility is desirable without the need of high tensile strength – the annealing treatment results in a microstructure with graphite nodules present in a fully ferritic matrix. For better machinability, carbide-forming elements should be as low as possible. There are three types of annealing treatments for different kinds of S.G. iron.

TABLE 14.4
Temperature Range of Stress Relieving for Various Types of S.G. Iron

Type of S.G. Iron	Unalloyed	Austenitic	Low-Alloy	High-Alloy
Temperature range (°C)	510–565	620–675	565–595	595–650

S.G., spheroidal graphite.

Heat Treatment of Cast Iron

- Subcritical annealing – This treatment converts pearlite to the ferritic phase. The presence of the carbide phase may result in a loss of ductility. For this treatment, the temperature range is 705°C–720°C. After heating and soaking, the castings are furnace cooled, followed by air cooling.
- Full annealing for unalloyed irons – This treatment is suitable for S.G iron with no alloying elements and eutectic carbide present. The temperature range is 870°C–900°C. After heating, the castings are furnace cooled till 350°C and then cooled in the air.
- Full annealing for S.G. iron–containing eutectic carbides – This treatment is given for castings having eutectic carbides. For heavier sections, larger soaking time is required, so that uniform temperature is achieved. After soaking, the castings are furnace cooled to about 700°C and then held for 2 h, followed by furnace cooling to 350°C and finally air cooling to room temperature.

Normalizing is another heat treatment process of heating S.G. iron castings to a temperature range of 870°C–940°C, holding for 1 h followed by air cooling. If alloying elements are present, then longer holding time is required. As a result of normalizing, tensile strength increases because of the formation of uniform and homogenous pearlite, and nodular graphite provided that the silicon content is low. The cooling rate depends on the weight of the casting specimen. Generally, more heavy castings shift the continuous cooling transformation curve toward the right so that the formation of pearlite and nodular graphite takes place even at a slow cooling rate. Tempering is done after normalizing in cases where martensite forms, which ensures to relieve the stresses due to the formation of the martensitic phase and get the desired hardness levels, tensile strength, and impact strength. Tempering is done by heating the castings to a temperature of around 400°C–600°C, soaking for 1 h followed by air cooling.

Hardening is the process of heating the cast to a temperature of austenitization temperature of 845°C–925°C. As the austenitization temperature increases, the hardness of the cast increases to the highest values up to 870°C, after which it starts decreasing due to the presence of a large amount of retained austenite. Oil quenching is generally preferred to avoid crack formation. Tempering is done immediately after the hardening or quenching process to relieve the stresses formed during the process at a temperature range of 420°C–600°C for at least a 1-h duration. The soaking time depends on the cast dimensions. The hardness, strength, and ductility of the tempered cast iron depend on the soaking time, composition, and temperature. The tempering takes place in two stages. The first stage consists of the precipitation of carbide phases followed by the second phase, which includes the nucleation followed by growth of the graphite nodules. This leads to a decrease in hardness as well as tensile strength.

The process of austempering for S.G. iron is similar to that of the austempering process of steel. The method includes heating the ductile iron to an austenitization temperature, quenching rapidly to temperature before the M_s temperature, holding it for a while, and cooling to room temperature. Usually, after austempering in steel, the bainite phase is formed, which is undesirable due to its low ductility and toughness. However, in ductile iron, acicular ferrite is uniformly dispersed in a carbon-rich metastable phase of austenite along with graphite nodules; this leads to good ductility as compared to other heat treatment processes.

Surface hardening treatment is done to improve surface properties like hardness and wear resistance. The heating time is short in these methods with little holding times; hence, pearlitic matrix is preferred. Caution is required in surface hardening of pearlitic irons to avoid crack formation. The cooling rate plays a vital role in these treatments. Water quenching of pearlitic cast irons has higher hardness than cast iron with the ferritic matrix. A fully ferritic matrix can form martensite only near the graphitic nodules, resulting in lower hardness values. Various parameters control the microstructure of the cast iron, such as temperature, time of holding, composition, and rate of cooling. Polymer quenchants can be preferred instead of water quenching to avoid the quench crack formation. The amount of alteration in the surface properties of the cast iron depends on the pearlite content. S.G. iron, which has already been hardened and tempered, contains secondary graphite that can act as a source to supply carbon during the surface hardening even for cast irons with ferritic matrix up to 90%–95% extent.

As already discussed, nitriding is a type of case hardening process that involves the diffusion process of nitrogen at a high temperature in the range of 550°C–650°C in the presence of an ammonia atmosphere. The treatment, in this case, leads to the formation of a thin layer with a high hardness value of 1100 VPN. The presence of alloying elements can increase the extent of case hardening. The nitriding process is done to improve the hardness, wear resistance, and corrosion resistance.

REFERENCE

1. Radzikowska, J. M. Effect of specimen preparation on evaluation of cast iron microstructures. *Mater. Charact.* **54**, 287–304 (2005).

FURTHER READING

Avner. *Introduction to Physical Metallurgy.* (Tata McGraw-Hill Education, New York, 1997).
Rajan, T. V., Sharma, C. P. & Sharma, A. *Heat Treatment: Principles and Techniques.* (PHI Learning Pvt. Ltd., New Delhi, 2011).
Singh, V. *Heat Treatment of Metals.* (Standard Publishers Distributors, New Delhi, 2006).

15 Heat Treatment Defects and Their Determination

Heat treatment plays an important role in improving the performance of steel by significantly enhancing its mechanical properties, leading to a longer life span. It can be described as the heating and cooling process that aims at improving the microstructure as well as the mechanical properties of metals and alloys. Various mechanical components such as gears and crankshafts are subjected to continuous friction as well as dynamic load. This requires the surface to possess high hardness and wear resistance, while the core should have sufficient toughness. Despite the adoption of requisite procedures of heat treatment, the heat-treated components may exhibit undesired and detrimental properties that can seriously affect the quality of the component in operation. If the latter occurs, then the material is said to be defective, and such kind of defects are known as heat treatment defects.

Heat treatment defects may also arise due to unsuitable chemical composition of the alloy and may pose a great threat during its operation, as the component may fail anytime without any prior indication. Therefore, it is necessary to know the possible causes of the heat treatment defects, so that it can be prevented on time. Inherent defects in the material, inadequate design of the tool, lack of precise control over furnace temperature and cooling/heating rates, and wrong selection of the material may be some of the causes of heat treatment defects. Thus, preventive measures such as proper selection of the defect free material, appropriate design of heat treatment cycle, improved automation, and adequate tools will be beneficial to get the optimized properties in the material after heat treatment.

The common types of defects that are usually observed in the heat treated components are as follows:

- Distortion
- Warping
- Residual stress
- Quench cracking
- Soft spots
- Oxidation and decarburization
- Low hardness and strength after hardening
- Overheating
- Burning
- Black fracture
- Deformation and volume changes after hardening
- Excessive hardness after tempering
- Corrosion and erosion

15.1 DISTORTION

The change in the shape and size (or dimensional change) of the heat-treated component due to thermal and structural stresses is known as distortion. Distortion can be classified as follows:

- Size distortion: change in volume. Mainly takes place due to expansion or contraction in the component.
- Shape distortion: change in geometrical form without any volumetric change. This is manifested by change in curvature such as twisting or bending in the component.

Distortion mainly occurs due to the following:

- Distortion in steel mainly occurs during hardening and tempering.
- Combined effect of thermal and transformation stresses causes distortion.
- It is also influenced by the heating rate, size, shape, thickness of the wall, and geometry of the component. Chemical and structural inhomogeneity, cooling rate, and subzero treatment are also certain influencing parameters.
- The presence of residual stress in the component before heat treatment may also cause distortion.

The risk or chances of distortion can be controlled by certain parameters that are as follows:

- Design: Sharp corners or sharp projections and thin walls should be avoided in the component.
- Composition: Proper selection of steel component in terms of its composition helps in the minimization of distortion.
- Uniform microstructure of the component and uniform temperature in the furnace helps in the minimization of distortion.
- Machining helps in reducing the risk of distortion.
- Subcritical annealing or normalizing operation helps in relieving the internal stresses and, thus, reduces the occurrence of distortion.
- Heating rate has to be very fast.
- Preheating reduces shape distortion in steels by reducing the thermal stresses produced due to the temperature gradient between the surface and the core.

15.2 WARPING

Warping can be defined as the asymmetrical deformation of heat-treated component during quenching. Nonuniform heating or cooling may lead to volumetric change, which further causes warping. Inadequate support of the work piece at high temperatures, presence of internal stresses, and inclination of the component in the quench bath are some of the causes of warping. Figure 15.1 shows a typical warping defect due to galvanization (submersed in a liquid zinc bath at around 450°C).

Heat Treatment Defects

FIGURE 15.1 Warping steel plates due to galvanization.[1]

The chances of warping can be minimized by the following:

- Addition of more alloying elements in the steel. This shifts the nose of the continuous cooling temperature (CCT) curve toward the right side, thus permitting a slow cooling rate. This reduces the generation of internal stresses, and thus, chance of warping gets reduced.
- Slow cooling in the martensitic range.
- Appropriate surface hardening treatments also may reduce warping.
- Annealing, normalizing, and tempering at high temperature before hardening also decrease the risk of warping.
- Uniform heating and quenching as well as proper alignment of the component in quenching bath also acts as a corrective measure for warping.

15.3 RESIDUAL STRESSES

Stress may be produced during transformation of austenite to martensite. Mainly, residual stresses arise due to the temperature gradient and uneven transformation across the cross section of a component. Residual stress can be minimized by the following:

- Slow down the cooling process so that temperature gradient can be lowered, and thus, transformation occurs at the same time throughout the cross section. Thus, it reduces the residual stress generation.
- Addition of more alloying elements in the steel shifts the nose of the CCT curve toward the right side, thus permitting a slow cooling rate.
- Residual stress can also be eliminated by stress-relieving process such as reheating steel to a relatively low temperature, around 250°C with a suitable holding time depending on the thickness followed by controlled cooling. At this temperature, 80%–85% of residual stress can be eliminated.
- Tempering at high temperature can also relieve the residual stress.

15.4 QUENCH CRACKING

Quench cracks generally appear in zigzag form at the grain boundaries, and they may be external or internal as well as small or large. They appear as straight lines that run from the surface toward the center of the quenched specimen as shown in

FIGURE 15.2 A quench crack extending from the surface to the core.[2]

Figure 15.2. This may be formed due to the presence of stresses produced during the transformation of austenite to martensite. They often appear after a steel sample undergoes quenching. The differential cooling in the surface and core during martensitic transformation is accompanied by the volume mismatch, which results in the generation of compressive stresses and thus, causes cracking. Generation of quench crack is detrimental, as the steel with quench cracks cannot be used further and, therefore, has only scrap value.

The preventive measures that can be followed include the following:

- Avoid manufacturing tools with sharp corners or sharp projections.
- There should be no residual stress, and if there is any, it should be relieved by annealing.
- Cool the component slowly in the martensitic range.
- Can follow the route of interrupted quenching, oil quenching, austempering, and martempering.
- Immediately temper after the quenching process to avoid quench crack formation.

15.5 SOFT SPOTS

When the hardness of the component surface is not uniform, even after the hardening process, it may lead to varying hardness at different points on the surface of hardened steels. Such kind of a defect is known as soft spots. The microstructure of such soft spot seems as dark region in white martensitic matrix. The various reasons pertaining to soft spots are as follows:

- Localized decarburization of steel
- Formation of vapor blanket between the quenchant and component during quenching, which hinders rapid heat dissipation
- Inhomogeneity of microstructure

Heat Treatment Defects

- Presence of foreign particles on the surface such as scales and dirt
- Heating of large components in furnaces, which may lead to nonuniformity in heating of the component
- Improper handling of component during quenching

Soft spot formation can be prevented by adopting spray quenching over conventional quenching.

15.6 OXIDATION AND DECARBURIZATION

Furnace atmospheric gases such as oxygen, carbon dioxide, and water vapor may react with the steel surface, whereas the latter is heated at a high temperature in an open furnace (open to the atmosphere). As a result, it gives rise to two surface phenomena: oxidation and decarburization.

Oxidation of the steel occurs when it reacts with oxygen, carbon dioxide, or water vapor. The possible series of chemical reaction are as follows:

$$2Fe + O_2 = 2FeO$$
$$4FeO + O_2 = 2Fe_2O_3$$
$$Fe + CO_2 = FeO + CO$$
$$3FeO + CO_2 = Fe_3O_4 + CO$$
$$Fe + H_2O = FeO + H_2$$
$$3FeO + H_2O = Fe_3O_4 + H_2$$

The gases also react with each other, and such kind of a reaction is known as water gas reaction.

$$CO + H_2O = CO_2 + H_2$$

The equilibrium relationship of iron and iron oxide with the gases at the operating temperature determines the degree of oxidation. When there is a high content of CO in a low content of water vapor in the furnace atmosphere, it leads to another reaction known as producer gas reaction as mentioned in the following:

$$2CO = CO_2 + C$$

Steel behaves like a catalyst in the producer gas reaction, and thus, oxidation of the steel surface occurs. At elevated temperatures (around 425°C), a porous and oxide layer grows, which leads to continuous disintegration of the component. This adversely affects the quality of the surface with consequential dimensional changes.

Decarburization involves the removal of carbon from the steel surface when it is heated at elevated temperatures (around 650°C). Carbon reacts with oxygen or hydrogen at this temperature, and this depth of carburization is a function of time, temperature, and furnace atmosphere. As shown in Figure 15.3, there is a

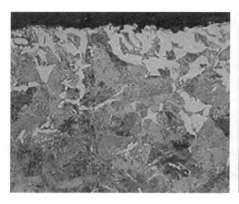 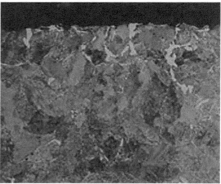

FIGURE 15.3 Decarburized surface showing a variation in the amount and depth of ferrite in as-rolled eutectoid steel.[3]

variation in ferrite content on the surface of as-rolled eutectoid steel. This phenomenon gravely reduces the quenched hardness, wear resistance, and fatigue strength. Decarburization involves the following reactions:

$$C + O_2 = CO_2$$
$$Fe_3C + O_2 = 3Fe + CO_2$$
$$C + CO_2 = 2CO$$
$$Fe_3C + CO_2 = 3Fe + 2CO$$
$$Fe_3C + H_2O = 3Fe + H_2 + CO$$

The measures that can be followed to reduce decarburization and oxidation include the following:

- Steel components can be heated in molten salts, vacuum, or controlled atmosphere.
- Machining can be done to remove the decarburized surface layer.
- Steel components may be heated with carburizing agents with surface coatings like borax.
- Components may be coated with ceramic material before the process of heat treatment.

15.7 LOW HARDNESS AND STRENGTH AFTER HARDENING

Sometimes, when a material or component is heated till the austenitizing temperature and then quenched, it may not develop the desired hardness and strength. The hardness is imparted by the martensite (hard phase) formation upon quenching. But it requires a further heat treatment, i.e., tempering to reduce or eliminate the brittleness of the as-quenched martensite. Tempering causes softening of the hard phase to enhance the other mechanical properties such as toughness, formability,

Heat Treatment Defects

and machinability. This greatly reduces the chance of crack nucleation. But there are certain reasons that may lead to low hardness and strength after quenching, such as the following:

- Lower hardening temperature and insufficient soaking time that may lead to the formation of products like bainite and pearlite, which have lower hardness value than that of martensite.
- Delayed quenching and slow cooling rate may also result the same as the latter.
- Higher tempering temperature may induce unwanted excessive softness in the component.
- Presence of large amount of retained (or untransformed) austenite, which is much softer than the martensite.

In such kind of a situation, subzero treatment can be carried out at around −30°C to −120°C, as the M_f transformation temperature may be below the room temperature.

Furthermore, improper surface hardening treatments like carburization, cyaniding, and nitriding process may lead to reduced hardness. Thus, improper carburizing atmosphere and temperature causes the aforementioned phenomena, and it can be overcome by considering proper carburizing and postcarburizing heat treatment.

15.8 OVERHEATING OF STEEL

Overheating involves the heating of steel above the upper critical temperature for a long period of time, which leads to coarsening of the grains. It adversely affects the mechanical properties and may lead to premature failing of gears due to fatigue. The optical and SEM images of a boiler steel tube are provided in Figures 15.4 and 15.5. The cooler side, which experiences relatively lower temperature, shows conventional lamellar pearlite structure. However, upon short-term excursion to high temperature (~750°C), distortion of the pearlitic structure results, leading to its spherodization resulting in poor strength.[4]

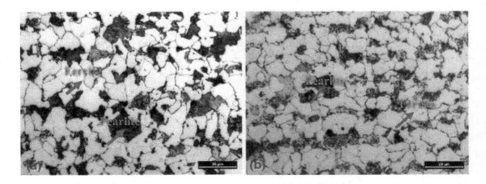

FIGURE 15.4 Optical microscope images of a boiler steel tube at (a) cooler side and (b) hotter side.[4]

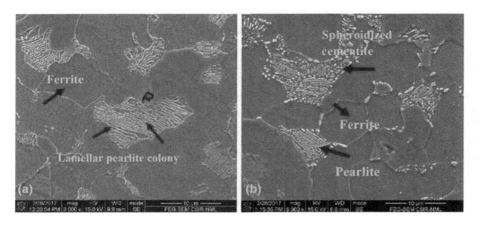

FIGURE 15.5 SEM images of a boiler steel tube at (a) cooler side and (b) hotter side.[4]

Effects of overheating are summarized as follows:

- Coarse-grained microstructure formation
- Widmanstätten structure formation in annealed steel
- Reduction in toughness and ductility
- Reduced hardness and coarse-grained martensite formation in the case of hardened steels
- Scale formation and surface decarburization
- Segregation of impurities like sulfide inclusions along grain boundaries, thus causing damage to the component

On the basis of composition, high-carbon and low-alloy steels, Ni–Cr–Mo, and high-alloy steels are more prone to overheating as compared to mild steel. On the basis of melting techniques, basic electric steels are more susceptible to overheating than the steels manufactured by open hearth process. Steels with higher inclusion content have higher overheating temperatures.

Overheated steels can be recovered by repeated normalizing process at temperatures 60°C–100°C higher than the conventional temperature. Repeated hardening by oil and tempering after prolonged soaking at 950°C–1120°C in carburizing atmosphere may also help in recovering the steel from its overheated condition. But the disadvantages of both these methods include dimension alteration and uneconomical process.

15.9 BURNING OF STEEL

If the temperature exceeds the solidus temperature while heating the steel, liquation may occur at the grain boundaries. In presence of an oxidizing condition, oxides may form and penetrate the grain boundaries. This phenomenon is known as burning of the steel, and its effect is permanent. Burnt steel becomes scrap. Effects of burning are summarized as follows:

Heat Treatment Defects 205

- Phosphorous gets segregated at the grain boundaries, and it subsequently gets precipitated as iron phosphide. This causes embrittlement of steels. The same kind of embrittlement may occur in the case of sulfur as well.
- High sulfur content has the ability to enhance both overheating and burning of steels.
- The components that are burned are unserviceable and cannot be reclaimed. It can hardly be used as a scrap.

On the basis of melting techniques, basic electric steels are more susceptible to burning than the steels manufactured by open hearth process.

15.10 BLACK FRACTURE

Black fracture can be characterized as carbon (specifically, graphite) inclusions in the steel, which renders undesirable properties to the steel. This is mainly caused due to excessive heating time and slow cooling after annealing. The corrective measure of such kind of a defect includes heating of the steel to a high temperature and forging it thoroughly and uniformly.

15.11 DEFORMATION AND VOLUME CHANGES AFTER HARDENING

Higher hardenability of the steels after the hardening process may result in severe deformation as well as undesirable volumetric changes. Such kinds of defects are detrimental to the mechanical properties of the component. Volumetric change, i.e., increase in volume, mainly occurs due to the martensitic transformation. The deformation is more severe in case of higher hardenability of the steels. Uneven heating, rapid cooling, improper support of the component in the furnace, improper dipping in the quenching bath, and presence of residual stresses may also result in deformation and volumetric changes.

Certain measures that can be considered to prevent such kind of a defect include the following:

- Use alloy steel that can undergo slight deformation by hardening.
- Cool the component slowly in the martensitic range.
- Apply surface-hardening treatments whenever possible.
- Perform stress-relieving treatment to reduce the residual stress.

15.12 EXCESSIVE HARDNESS AFTER TEMPERING

Insufficient or excessive holding time during tempering may lead to enhanced hardness or insufficient hardness, respectively. This may occur due to low temperature and/or insufficient soaking time while tempering. Maintenance of high temperature during tempering may also cause such kind of a defect. This defect may also arise if the hardening temperature is too low or if the cooling rate is too low.

Such kind of a defect can be prevented by the following:

- A second tempering treatment with proper temperature and soaking time
- Subsequent normalizing or annealing followed by hardening

15.13 CORROSION AND EROSION

Pitting-type corrosion of the component or localized corrosion may occur during heat treatment due to the presence of high content of sulfuric salts (0.7%–0.8%) in the molten salt bath. Additionally, when the bath is rich in oxygen or iron oxides, it causes pitting corrosion of the component. The same conditions may also cause erosion. Erosion can be defined as the reduction in size of the component due to the loss of material from its surface.

Thus, its preventive measures are as follows:

- Controlling the salt composition in the bath
- Deoxidizing the bath

REFERENCES

1. Association, A. G. Distortion & Warping. American Galvanizers Association https://galvanizeit.org/design-and-fabrication/design-considerations/distortion-and-warping (2020).
2. 7 Causes for Quench Cracking of Steel. Speaking of Precision Blog https://pmpaspeakingofprecision.com/2010/08/03/7-causes-for-quench-cracking-of-steel/ (2010).
3. Vander Voort, G. F. *Understanding and Measuring Decarburization.* (ASM INT Subscriptions Specialist Customer Service, Materials Park, OH, 2015).
4. Munda, P., Husain, Md. M., Rajinikanth, V., & Metya, A. K. Evolution of microstructure during short-term overheating failure of a boiler water wall tube made of carbon steel. *J. Fail. Anal. Prev.* **18**, 199–211 (2018).

FURTHER READING

Rajan, T. V., Sharma, C. P., & Sharma, A. *Heat Treatment: Principles and Techniques.* (PHI Learning Pvt. Ltd., New Delhi, 2011).

Sinha, A. K. Defects and distortion in heat-treated parts. *ASM Handb.* **4**, 601–619 (1991).

Singh, V. *Heat Treatment of Metals.* (Standard Publishers Distributors, New Delhi, 2006).

16 Some Special Heat Treatment Practices

In present days, steel is an irreplaceable material being used in a wide spectrum of applications. However, the requirements for various applications may differ in a gross scale. Hence, appropriate selection of steel is the key to achieve the desirable in-service performance. It is well established, at this point of time, that heat treatment of steel plays a vital role in altering its properties. Hence, a well-planned heat treatment cycle must be designed in order to comply with the stringent application requirements. In this chapter, a brief knowledge has been shared about the type of steels and respective heat treatment operations carried out by different industries.

16.1 AUTOMOBILE INDUSTRIES

Heat treatment is a key technology to enhance the operative use of materials and to achieve the desirable combination of properties used in the automobile components. The manufacture of automobiles involves the utilization of a variety of materials and technologies. Vehicle weight reduction via the use of high-strength and light-weight structural materials is a necessity in the present day as it aims towards energy conservation. Though, aluminum and plastics have seen an increase in use, ferrous alloys constitute about 70% of a current vehicle. Some of the steel components used in automobiles are shown in Figure 16.1. Metallic materials are preferably matched for specific applications where the material is expected to work in a very reliable way even under heavy stresses. Effective heat treatment is necessary for optimizing properties of nearly all types of metallic components with durability being a feature of prominence in a great number of applications. Starting with raw metal products to final component assembly, with various stages of processing, several types of heat treatments are performed in the manufacturing of different types of automotive components. Appropriate heat treatment processes render the required combination strength and/or hardness with toughness, which is dictated by the respective components' applications.

Figure 16.1 shows the various automotive components utilizing iron and steel as the building material. The 70% steel components used in automobiles include the drive shafts, camshafts, crankshafts, cylinder heads, bearings, valves/outlets, rocker arms, suspensions, gears, clutch parts and many more. Specific heat treatments for parts are as follows:

16.1.1 DRIVESHAFTS/JOINTS

Driveshafts are the automotive components where safety has top priority as it functions to transfer torque from the gearbox to the wheels (Figure 16.2). Heat treatment

207

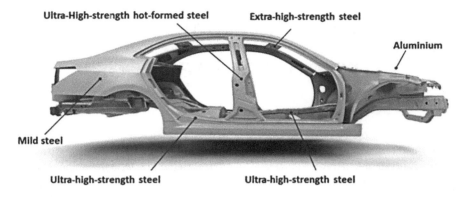

FIGURE 16.1 Some components of a vehicle.[1]

FIGURE 16.2 Driveshafts as used in automobile applications.[2]

is necessary to improve service life of the component and to counteract wear. Case-hardening treatments like vacuum hardening (vacuum carburizing) or gas nitriding as well as plasma nitriding are suitable.

16.1.2 Suspension

Suspensions connect each wheel to the body of the vehicle (Figure 16.3). Shock absorbers, handlebars, wheel carriers, joints and springs are the various components of the suspension system. To ensure longevity of these components case-hardening treatments like inert gas hardening and vacuum hardening are the heat treatments performed.

16.1.3 Clutch

Clutch acts as the link between the power train and the drive train (Figure 16.4) and thereby, plays a crucial role in the transfer of power and torque from the engine to the wheels. Clutch plates are subjected to high wear stresses over frequent gear changes

Some Special Heat Treatment Practices

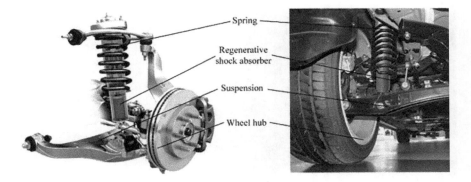

FIGURE 16.3 Suspension shock absorber used in automobile applications.[3]

FIGURE 16.4 Clutch as used in automobile applications.[4]

and are the most likely to be worn out the earliest. Good strength and toughness are vital attributes given the stresses to which these components are subjected to. Plasma nitriding is a suitable heat treatment process.

16.1.4 Crank Shafts

Crank shafts are responsible for the conversion of linear motion of pistons to rotary motion (Figure 16.5). It ensures that this component is subject to tensile, compressive, and radial forces. Sufficient bending and torsional strength of the shaft and good loading capacity of the bearing points thus becomes a necessity. Gas nitriding, nitrocarburizing, plasma nitriding, and case hardening are suitable thermochemical processes. Nitrogen addition leads to increase in strength and development of compressive forces. This results in increased load-bearing capacity. Carburization of surface layer is chiefly used in racing cars.

FIGURE 16.5 Crank shaft as used in automobile applications.[5]

16.1.5 Bearings

Bearings transmit forces and movement between machine parts, as they slide or roll over each other. These components are subjected to high loads. To limit resistance generated during loading gas nitriding, plasma nitriding, case hardening, and vacuum hardening are the treatment used to increase the strength.

16.1.6 Gearwheels/Planetary Gears

Gearwheels (Figure 16.6) are used to transmit torque. Gears stand a chance to shatter under extreme loads. In order to show elastic behavior throughout life cycle and withstand load shocks surface strength and inner toughness of the gears are required. Case hardening and nitriding treatments improve strength and wear behavior.

The various types of heat treatments used to treat automotive parts are given in Table 16.1.

16.2 AEROSPACE INDUSTRIES

Heat treatment process plays an important and is frequently used to alter the properties of materials used for aerospace industries. Aerospace industry requires high quality and precise dimension requirements for the components used. Materials used in this sector are required to have high strength along with good fatigue life. Stress reduction is required for achieving appreciable strength and fatigue life. This will help in ensuring that the materials tolerate extreme temperature and pressure conditions without failing, also precise and specific dimension requirements for the materials and components are there, which requires the materials to sustain these extreme conditions without changes in their shape or size. Heat treatment proves to be an essential step in this regard.

Some Special Heat Treatment Practices

FIGURE 16.6 Schematic presentation of gears as used in automobile applications.[6]

TABLE 16.1
Various Automotive Components and Respective Heat Treatment Processes[7]

Types of Heat Treatment	Purpose	Typical Components
Annealing	Softening and removing residual stress and imparting substantial ductility for post processes	Forged blanks for gearing and miscellaneous parts
Quenching and tempering	Optimized combination of hardness for strength and toughness	Fasteners, rods, and arms
Austempering	Optimize microstructure and hardness via isothermal transformation	Cast iron brackets, high-carbon springs
Case hardening: Carburizing Nitriding Carbonitriding Nitro-carburizing Oxy-nitro-carburizing	For improving fatigue strength and wear resistance	Gears and shafts, oil pump gears, brake pad liner plates, cam shafts

Commonly heat-treated components for aerospace industries include engines, frame parts, cooling systems, and hydraulic fittings. Heat treatment processes and techniques include a number of methods that improve the properties of the material used in this industry namely, metals and alloys along with extending the life of these

212   Phase Transformations and Heat Treatments of Steels

components. These treatments along with the aerospace industry can be used also in aviation, commercial, and business aircraft carriers and helicopters. Different heat treatments can be given depending on the metals and alloys used and according to the final property requirement of the end product. Aerospace industry is a vital field of research where new engine designs, various R&D process, new materials performance research along with testing of the joining systems, and novel materials are highly dependent on the production technique and the post processing methods given to it. Controlled atmosphere heat treatments like case hardening, tempering, etc. are commonly done for components like flap tracks, landing gear parts, and airframe during the manufacturing process. The nitriding of engine components such as gears leads to improved and desirable properties suiting to aerospace industry. The key heat treatment processes are as follows:

i. *Vacuum heat treatment*
   a. Ageing
   b. Annealing
   c. Homogenizing
   d. Low pressure carburizing
   e. Plasma nitriding
ii. *Controlled atmosphere treatment*
   a. Carbonitriding
   b. Carburizing
   c. Case hardening
   d. Gas nitriding
   e. Hardening and tempering
   f. Nitrocarburizing

In aerospace industry, safety of the passengers and the pilot is utmost important for which the components should be reliable and as per the safety regulations. Hence, reliable operations of the components are given to components to avoid fatal results. Following are the components where heat treatment is commonly done.

16.2.1 SEAT EJECTORS

Ejector seats (Figure 16.7) for fighter aircrafts are manufactured to be as precise and as close to the intended properties as possible. Heat treatment of these ensures that this essential equipment operates properly without any malfunctioning. A countless number of lives of pilots are saved and will continue to be saved by the efficient deployment of these seats in case of any uncontrolled flights and accidents. Seat ejectors are small yet an important part of these aircrafts.

16.2.2 TURBINE BLADES

Aircraft blades (Figure 16.8) are subjected to extreme temperatures and conditions while they are in operation. These components operate frequently at temperatures tending to their melting point and hence can cause failure of these components while

Some Special Heat Treatment Practices

FIGURE 16.7 Seat ejectors as used in aircrafts.[8]

FIGURE 16.8 Turbine blades as used in engines of aircrafts.[9]

in use. Hence, to avoid early failure, heat treatment operations along with other suitable methods and techniques are prominently used allowing these turbine blades to withstand these high temperatures for prolonged period of time.

16.2.3 Landing Gear

Safety components like aircraft landing gear (Figure 16.9) are expected to perform well without any fault and should withstand environmental effects and avoid failure mechanisms like fatigues every time the aircraft takes a flight. Several combinations of heat treatment processing techniques are given to ensure the properties of the steel are optimized which can help to protect it during the service life. Landing gears were used to be surface-treated using hard chrome plates which are now replaced by other environmentally friendly processes and thermal spraying techniques. These help the landing gears to withstand wear and corrosion during their working life.

16.3 MEDICAL EQUIPMENT

Various medical equipment and supplies namely medical implants, dental tools, surgical devices, and so on are frequently heat-treated and given thermal processing to tailor the required specific properties of these equipment. Commonly used material for these medical equipment includes stainless steel which can be given ideal heat treatments from plenty of other available options. Medical industry demands in

FIGURE 16.9 Landing gear in aircrafts.[10]

Some Special Heat Treatment Practices

choosing the right material with specific properties like good mechanical strength and corrosion resistance along with biocompatibility. Biocompatibility is extremely essential when medical industry is involved. Material lacking in this property can lead to compromise of the life and health of people rendering these materials. Other metals, namely, titanium are also used for hip implants that need to be given special treatments to achieve stringent properties and proper finish.

16.3.1 SURGICAL EQUIPMENT

Surgical equipment falls under the category of nonimplant medical devices. This also includes dental instruments, surgical staples, guide pins, needs, trays, and storage cabinets. The materials used in these applications should have good corrosion resistance, good mechanical properties, formability, and biocompatibility. Along with biocompatibility, surgical equipment also need good corrosion resistance to withstand the attack of bodily fluids and inhibit reactions avoiding formation of harmful product. Surgical equipment are typically made from 420 stainless steel grade which is a high-carbon steel that can be heat-treated leading to hardening. This material exhibits excellent corrosion resistance properties upon hardening and offers appreciable ductility upon annealing condition. In hardened state, it can withstand water, alkali, and mild acid attacks; however, in annealed state, the corrosion resistance tends to decrease.

16.3.2 MEDICAL DEVICE FASTENERS

Fasteners are used in a number of applications in the medical industry namely in the dental and orthopedic implants. They help to keep all the different components and parts of the implant together in an assembly. These components and parts have different shapes and sizes. Usually, these components in medical industries are small and tiny, but failure of a fastener will result in failure of the whole implant. This needs the fastener to be suitable, precise, and compatible with other equipment and parts along with good mechanical properties and biocompatibility. Only then, the fastener can be expected to perform as desired. Choice of suitable fastener can cut costs of the whole implant, improving the overall quality of the final product.

16.3.3 IMPLANTS

Surgical implants are often made from specific grades of stainless steel, i.e. austenitic type. Examples are bone plates, screws, femoral fixation device, clips, nails, and pins. Titanium is also gaining major interest of the researchers for use in implants along with cobalt-based alloys. For dental and hip implants, chemical bonding between the implant and the body/bone does not occur. This is because of the lack of formation of chemical bond between them. Hence, the surface of the implant is made with high porosity such that bone tissues can grow into the pores leading to a stable fixture and integration. This is done by surface modification of the implant, which creates a good connection between the implant and the body. Various metals used for different implants are tabulated in Table 16.2.

216   Phase Transformations and Heat Treatments of Steels

TABLE 16.2
Various Metals Used in Different Implants[11]

Type of Implant	Specific Use	Metal Used
Cardiovascular	Stents, artificial wires	316L SS, CoCr, Ti
Otorhinology	Artificial eardrum	316L SS
Craniofacial	Plate and screw	316L SS, CoCrMo, Ti
Orthopaedic	Bone fixation, artificial joints	316L SS, Ti, CoCrMo
Dentistry	Orthodontic wire, fillinf	316L SS, CoCrMo, TiMo

16.4 DEFENSE INDUSTRIES

The standards of materials used for mission-critical applications in the defense industry are very high. When working with a metallic material, precision and quality are considered two of the most important factors for manufacturers in the defense industries. Before adding to the defense equipment, it is essential that the metal parts be treated correctly. Tanks, fighter jets, stealth bombers, and helicopters all need precision-treated metallic components.

16.4.1 ARMOR OF COMBAT VEHICLES

The armor of combat vehicles needs to be light in weight and has great ballistic protection. Ultra-high strength (UHS) steel is the most preferred metallic armor used today due to the properties of good toughness with high strength and hardness and ease of heat treatment that they possess. Quenching and tempering produce strengthening and increase impact toughness of UHS steel. Martensitic structure formed during quenching has the highest strength, but internal stresses developed during transformation limit its use in untempered condition. To boost absorption of impact energy, the ductility and toughness of material need to be increased. Tempering serves this purpose.

Austenitization of UHS steel is carried out at 910°C for 1h followed by oil or water quenching. Tempering up to 600°C is done, with different tempering temperatures producing varying properties. Table 16.3 gives the values of various mechanical properties of UHS steel tempered at different temperatures after quenching.

From the varying mechanical properties, suitable tempering temperature for specific applications may be used.

16.4.2 FIREARMS

Different types of firearms are used for various purpose nowadays. Warping of barrels is a problem in rifles, pistols, and shotguns. Warping occurs due to continuous stress during constant firing that heats up the barrel. Heat treatment processes such as case hardening, tempering, gas nitriding, and surface hardening increase performance and accuracy of firearms. Maintenance and cleaning becomes easier as well.

Some Special Heat Treatment Practices

TABLE 16.3
Effects of Tempering Temperature on the Mechanical and Ballistic Properties[12]

Tempering Temperature (°C)	Yield Strength (MPa)	UTS (MPa)	Impact YS/UTS (VHN)	Hardness (J)	%Reduction (mm)	%Elongation
0	1367	1900	0.72	586	44	9
200	1417	1808	0.78	555	48	12
300	1463	1700	0.86	518	43	7.5
400	1433	1587	0.90	490	48	10
500	1286	1409	0.91	442	53	12
600	1146	1247	0.92	381	60	16

16.4.3 MISSILES

Missiles are normally made of materials having very high resistance to environmental conditions. Thermal treatment is necessary in the manufacturing of missile systems. All the components starting from the mounting hardware to the guidance systems require treatment. Hardening, induction heat treatment and age hardening are treatments that can be performed on missile motor cases, wing fins, mounting hardware, and guidance components.

REFERENCES

1. Asnafi, N., Shams, T., Aspenberg, D. & Öberg, C. 3D metal printing from an industrial perspective—product design, production, and business models. *BHM Berg-Hüttenmänn. Monatshefte* **164**, 91–100 (2019).
2. Driveshaft. Simple English Wikipedia, the free encyclopedia (2014).
3. Wang, Z., Zhang, T., Zhang, Z., Yuan, Y. & Liu, Y. A high-efficiency regenerative shock absorber considering twin ball screws transmissions for application in range-extended electric vehicles. *Energy Built Environ.* **1**, 36–49 (2020).
4. Dual-clutch transmission. *Wikipedia*. https://en.wikipedia.org/wiki/Dual-clutch_transmission.
5. Yu, Z. & Xu, X. Failure analysis of a diesel engine crankshaft. *Eng. Fail. Anal.* **12**, 487–495 (2005).
6. Qin, Z., Zhang, Q., Wu, Y.-T., Eizad, A. & Lyu, S.-K. Experimentally validated geometry modification simulation for improving noise performance of CVT gearbox for vehicles. *Int. J. Precis. Eng. Manuf.* **20**, 1969–1977 (2019).
7. Funatani, K. Heat treatment of automotive components: current status and future trends. *Trans. Indian Inst. Met.* **57**, 381–396 (2004).
8. Ejection seat. Wikipedia (2019).
9. El-Sayed, A. F. *Aircraft propulsion and gas turbine engines.* (CRC press, Cleveland, OH, 2017).
10. Landing gear. Wikipedia (2020).
11. Hermawan, H., Ramdan, D. & Djuansjah, J. R. P. Metals for biomedical applications. *Biomed. Eng. - Theory Appl.* (2011) doi:10.5772/19033.
12. Jena, P. K. et al. Effect of heat treatment on mechanical and ballistic properties of a high strength armour steel. *Int. J. Impact Eng.* **37**, 242–249 (2010).

Index

A

Age hardening, 173, 217
Ageing, 212
Allotropic transformation, 76, 80
Alloy carbides, 140, 144, 150–151, 166
Alloying elements, 86–89, 110, 113–115, 154, 157, 162, 165, 172–173
Annealing, 133–142, 167
Annealing twin, 48–49
Atomic bonding, 54
Atomic diameter, 47–48
Atomic number, 1–2
Atomic weight, 1
Ausforming, 165–166
Austempering, 151, 195
Austenite, 26–27, 83
Austenite stabilizer, 88, 169
Austenitic alloys, 88
Austenitic stainless steel, 171
Austenitization, 102, 193
Automobile industries, 179, 207

B

Bainite, 111–113
Bainitic transformation, 111
Bainitic zone, 111
Batch furnace, 118
Bearing, 209–210
Bell furnace, 121–122
Black fracture, 205
Body centred cubic (BCC), 6, 11
Bogie hearth furnace, 121
Box-type batch furnace, 119
Brass, 52
Bravais lattice, 6–10
Brine, 146–147
Burgers vector, 42–46
Burning of steel, 204

C

Carbide, 82–89
Carbonitriding, 162–163
Carburizing, 158–161
Case hardening, 158
Cast iron, 185
Cementite, 82–83
Climb, 46
Closed γ-field, 89
Close packed, 6, 11–12

Close Packed Hexagonal (CPH), 6, 12–13
Clustering, 61
Clutch, 208–209
Coarsening, 108, 149
Cold rolling, 1, 67, 177
Cold working, 138
Component, 211
Configuration, 2–4
Configurational entropy, 56
Continuous cooling transformation (CCT), 102
Continuous furnace, 123
Contracted γ-field, 90
Conveyor furnace, 123
Coordination number, 10–11, 30, 47
Corrosion, 206
Covalent bond, 4–5
Crank shafts, 209
Creep, 191
Critical cooling rate, 105, 141
Crystallographic direction, 47
Crystallographic plane, 32
Crystal structure, 5–6, 13–14
Crystal system, 6–10
Cubic, 9
Curie temperature, 80, 86

D

Degree of freedom, 65
Dendrite, 186–187
Density, 5, 115
Density of dislocation, 115, 161
Diffusion, 21
Diffusionless transformation, 113
Dislocation, 42–46
Dispersion, 141, 151
Distortion, 198
Driveshaft, 207–208
Dual-phase steel, 173
Ductile fracture, 154
Ductile iron, 190–191
Ductile tempered martensite, 182
Duplex stainless steel, 182–183

E

Edge dislocation, 42–43
Elastic strain energy, 46
Electrical conductivity, 4, 37
Electrical neutrality, 1, 3
Electrical resistance furnace, 118

219

220 Index

Electric arc, 185
Electric grade steel, 177
Electron to atom ratic, 53
Enthalpy of formation of a solid solution, 54
Entropy of formation of a solid solution, 56
Equilibrium, 63–65
Erosion, 206
Eutectic phase, 72–73
Eutectic phase diagram, 75
Eutectic transformation, 72–73
Eutectoid composition, 87–88, 102
Eutectoid steel, 103–104, 107–108
Eutectoid temperature, 85–87, 102
Expanded γ-field, 88–89

F

Face centred cubic (FCC), 10
Fatigue, 138, 153
Ferrite, 76, 85
Ferrite stabilizer, 88
Fick's first law, 24
Fick's second law, 25
Flame hardening, 163–164
Fracture, 154
Free energy, 28, 58
Full annealing, 134

G

Gamma iron, 80, 113
Gas carburizing, 130, 153
Gearwheels, 210
Gibb's Phase rule, 64
Glide, 44–45
Grain, 47
Grain boundary, 47–48
Grain growth, 108, 134
Grain size, 108
Graphite, 82
Graphite flake, 185
Graphitization, 188
Gray cast iron, 188
Grossman's critical diameter, 155
Growth kinetics, 98

H

Hardenability, 153
Hardness, 115, 141
Heat affected zone (HAZ), 179
Heterogeneous Nucleation, 95
Hexagonal, 6–7
Hexagonal Close Packed (HCP), 11, 49
High Strength Low Alloy (HSLA) steels, 172

Homogenization, 27, 108
Homogenous Nucleation, 91
Hot worked steels, 135
Hume-Rothery rules, 53, 66, 71
Hypereutectic, 73–74, 85
Hypereutectoid, 86, 134–135
Hypoeutectic, 72–74, 85, 186
Hypoeutectoid, 86–87, 134

I

Implants, 215
Inclusion, 157, 204
Induction furnace, 117–118
Induction hardening, 192
Intergranular corrosion, 169
Interlamellar spacing, 109
Intermediate phase, 61
Intermetallic, 173
Interstitial atom, 21–22
Interstitial diffusion, 21–22
Interstitial site, 51–52
Interstitial solid solution, 42, 51–52
Invariant, 68, 82
Isoforming, 165–167
Isomorphous, 66–67
Isothermal, 104, 106
Isotope, 1

J

Jominy curve, 157
Jominy end quench test, 156

L

Lamellar, 107, 111
Lath martensite, 115
Laves, 61–62
Ledeburite, 83, 186
Lever rule, 65
Liquidus, 68, 72–73
Lower bainite, 112

M

Malleable cast iron, 188
Manganese Steel, 172
Martensite, 113–115
Martensitic transformation, 113
Medical equipment, 214
Metallic bond, 5
Metastable, 82, 105
Mf temperature, 114
Microstructure, 64, 73, 85, 140, 200

Index

221

Mild steel, 164, 172, 204
Missiles, 217
Monoclinic, 7
Morphology, 109, 111, 164
Morphology of martensite, 115
Ms temperature, 114
Muffle furnace, 119

N

Nickel steels, 167–168
Nodular cast iron, 190
Nodular iron, 185
Nodules, 109–110, 189–190
Normalizing, 141
Nucleation, 91
Nucleation barrier, 94

O

Octahedral voids, 80–82
Open γ-field, 88
Order - disorder transformation, 76
Orthorhombic, 8–9
Overheating, 203
Oxidation, 201

P

Pack carburizing, 119
Packing factor/efficiency, 14
Partial annealing, 137
Partially soluble, 67, 75
Patenting, 139
Pearlite, 102–104
Pearlitic transformation, 108
Peritectic phase diagram, 75
Peritectic reaction, 83
Peritectoid reaction, 76
Phase diagram, 63
Phase rule, 64–65
Plate martensite, 115
Polymer quenchant, 145, 196
Precipitation, 100–101, 150, 172
Proeutectic, 73–74, 85

Q

Quenchant, 145, 147
Quench cracking, 199
Quenching, 145

R

Raoult's law, 71
Recovery, 137, 149

Recrystallization, 137, 138
Residual stresses, 199
Rotary hearth furnace, 123

S

Salt bath furnace, 125
Screw dislocation, 44
Seat ejector, 212
Secondary austenite, 183
Secondary graphite, 196
Secondary hardening, 150–151
Sensitization, 171
Sigmoidal curve, 110
Slip, 42–46
Solid solution, 51
Solidus line, 68
Spheroidal graphite (S.G.) iron, 190
Stabilized austenite, 115
Stainless steel, 169
Strain energy, 81
Strain hardening, 180
Strain-induced transformation, 172
Strain rate, 49
Stress- relieving annealing, 138
Stretcher-strain, 173
Substitutional solid solution, 52
Surface hardening, 158, 196
Surgical equipment, 215

T

Temperature-time-transformation (TTT)
    diagram, 102–106
Tempering, 148
Tetragonal, 9
Tetragonality, 113, 149
Tetrahedral, 14, 80–82
Thermomechanical treatment, 164
Tie line, 68–70
Tilt boundary, 48
Tool steels, 173
Toughness, 137, 148–151
TRIP (Transformed Induced Plasticity) steels,
    172
Tunnel furnace, 124
Turbine blade, 213
Twin boundary, 48
Twist boundary, 48

U

Upper Bainite, 111
Upper critical temperature, 86, 134, 137

Index

V

Vacancy, 21–22, 37
Van der Waals force, 5
Void, 40–41

W

White cast iron, 186
Work hardening, 172

Y

Yield point, 173
Yield strength, 79, 142, 172

Z

Zener's theory, 100